William Nicholson, C Pajot des Charmes

The Art of Bleaching Piece-Goods, Cottons, and Threads, of Every Description

Rendered more easy and general by means of the oxygenated muriatic acid; with the method of rendering painted or printed goods perfectly white or colourless

William Nicholson, C Pajot des Charmes

The Art of Bleaching Piece-Goods, Cottons, and Threads, of Every Description
Rendered more easy and general by means of the oxygenated muriatic acid; with the method of rendering painted or printed goods perfectly white or colourless

ISBN/EAN: 9783337393373

Printed in Europe, USA, Canada, Australia, Japan

Cover: Foto ©berggeist007 / pixelio.de

More available books at **www.hansebooks.com**

THE ART OF BLEACHING PIECE-GOODS, COTTONS, AND THREADS, OF EVERY DESCRIPTION,

Rendered more easy and general by Means of the Oxygenated Muriatic Acid; with the Method of rendering painted or printed Goods perfectly white or colourless. To which are added, the most certain Methods of bleaching Silk and Wool; and the Discoveries made by the Author in the Art of bleaching Paper.

ILLUSTRATED WITH NINE LARGE PLATES,

IN QUARTO,

REPRESENTING ALL THE UTENSILS AND DIFFERENT MANIPULATIONS OF THE BLEACHING PROCESS.

AN ELEMENTARY WORK,

COMPOSED FOR THE USE OF MANUFACTURERS, BLEACHERS, DYERS, CALLICO PRINTERS, AND PAPER-MAKERS.

BY PAJOT DES CHARMES,

FORMERLY INSPECTOR OF MANUFACTURES, MEMBER OF THE LYCEUM OF ARTS, OF THE SOCIETY OF INVENTIONS AND DISCOVERIES OF THE PHILOMATHIC SOCIETY, IN FRANCE.

TRANSLATED FROM THE FRENCH,

WITH AN APPENDIX.

LONDON:

PRINTED FOR G. G. AND J. ROBINSON, PATER-NOSTER-ROW.

1799.

CONTENTS.

CHAPTER I.

PRELIMINARY OBSERVATIONS,

BY THE AUTHOR.

THOUGH the illuſtrious Swediſh chemiſt Scheele was the firſt who obſerved the property of the oxygenated muriatic acid, which was alſo a diſcovery of his own, of diſcharging vegetable colours, has acquired the ſtrongeſt claim to the gratitude of the public, it is equally true, that the celebrated French chemiſt Berthollet has eſtabliſhed an equal claim to the acknowledgments of the world, by his active and able exertions on an object of ſo much conſequence to the commerce of the linen and cotton manufactures. The different memoirs which he has publiſhed on this ſubject, particularly that which is inſerted in the ſecond volume of the Annales de Chimie—the ſcientific application he has made of this acid to diſcolour the ſeveral vegetable ſubſtances which conſtitute the raw materials of manufactures—the particular developement, which the proſperity of thoſe manufactures led him to conſider as neceſſary to excite emulation among ſpeculators, ſoon afforded very promiſing reſults, as might naturally be expected from the publication of ſo uſeful a proceſs. Manufacturers in all parts of the nation were induced to conſult chemical and philoſophical men, in order to obtain information reſpecting it: the happy conſequences which have rewarded their labours in this reſpect, are truly honourable to the zeal of the parties themſelves, and have added to the reputation of their guide in this new department of reſearch.

The knowledge which I had acquired reſpecting the inconvenience and delay of the

common

common proceſs of bleaching—the incalcula-ble advantages which I ſaw muſt attend the method propoſed by Berthollet—the new life which the manufactures of thread and piece-goods, and the commercial tranſactions dependent thereon, would certainly receive:—theſe views, added to the deſire of knowledge, and a wiſh to contribute to the propagation of a diſcovery which promiſed an increaſe of our riches and our enjoyments, engaged me to verify the proceſs deſcribed in the Annales de Chimie. My intention was, in the firſt place, to make myſelf maſter of the proceſs, and then to propoſe, with confidence, this new method of bleaching to the manufacturers, merchants, and bleachers, in my department of inſpection, to whom this ſpecies of induſtry might prove advantageous, and to give them every information in my power. But I ſoon found that it would be in vain to deſcribe and publiſh this method, which would

would be in a ſhort time forgotten or confined to a few individuals, if it were not rendered more economical, leſs dangerous, and more amply deſcribed with regard to the manipulations, or practical part, ſo as to be rendered eaſy and perfectly adapted to the comprehenſion even of workmen totally unacquainted with chemical operations. For I knew that maſters have ſeldom the time, or will take the trouble, to operate themſelves, but moſt commonly confide their work to men upon whom they can depend. I therefore took the utmoſt pains to render the diſcovery of bleaching with the oxygenated muriatic acid of general uſe. As I had the good fortune to be ſucceſsful in my trials, I ſhall endeavour in the following work to deſcribe the proceſſes, by the aſſiſtance of which I ſucceeded.

I firſt give an account of the principal difficulties I experienced in operating, according to the

the directions in the memoir before-mentioned; difficulties which the author himſelf would not have failed to remove, if he had himſelf operated in the large way. I then deſcribe the methods which I have thought proper to ſubſtitute, inſtead of ſeveral of thoſe which are there pointed out: and, laſtly, I deſcribe, with the greateſt preciſion and minuteneſs, the different operations which are indiſpenſably neceſſary to give linen, hempen, cotton and mixed goods, a perfect bleaching, equal in colour to the beſt which are met with in the market, and are known in France by the names of the white goods of Troyes, Rouen, Senlis, &c.

ADVERTISEMENT

BY THE TRANSLATOR.

PHILOSOPHICAL men, as well as manufacturers, will, no doubt, receive with satisfaction the following treatiſe on a new art of great importance to ſociety, and truly honourable to philoſophical chemiſtry. It is unneceſſary to enlarge on the value of a proceſs, which has been eagerly and univerſally adopted as ſoon as ever it was known, and its principal difficulties removed. Neither will it require any argument to ſhew the advantage which practical men muſt receive from a detailed and very faithful account of proceſſes, conducted on a ſcale of ſufficient magnitude for commercial purpoſes.

 When

When we reflect on the ftate of thofe arts which are mentioned in the title page, and the numerous applications this new method is ftill capable of, as well as the utility of teaching it to all who are in any refpect concerned in bleaching, it will fcarcely be queftioned, but that this elementary treatife muft prove of great public benefit.

WILLIAM NICHOLSON.

Newman-ftreet, June 13, 1799.

THE ART

OF

BLEACHING COTTONS, THREAD, &c.

BY

THE OXYGENATED MURIATIC ACID,

RENDERED OF MORE EASY AND GENERAL USE *.

CHAP. I.

An Explanation of the Difficulties which attend the Method of Bleaching, deſcribed in the ſecond Volume of the Annales de Chimie, when practiſed by inexperienced Operators.

ONE of the firſt difficulties, in the operation illuſtrated by the plate annexed to the memoir in the Annals of Chemiſtry, is to obtain in the depart-

* It was intended that this work ſhould have been publiſhed, in 1791, by the General Adminiſtration of Commerce (in France); but the ſuppreſſion of that board, in the courſe of the ſame year, prevented its appearance. Since that time, other circumſtances have been unfavourable to the author's intention of publiſhing the reſults of his experiments on the new method of Bleaching.—*Note of the Author.*

departments ſuch furnaces as are there preſcribed to be uſed. They can ſcarcely be had but by ſending to Paris, where they are made: and if it be even ſuppoſed that theſe furnaces might conveniently be made at a diſtance from the capital, they would ſtill appear to be coſtly, ſuitable only to a ſingle matraſs, not calculated to ſhew the proceſs which takes place in the veſſel, ſubject to be heated too ſpeedily, and liable to render part of the lutes difficult to hold; laſtly, they do not always ſecure the operation from the conſequences of an abſorption of the water of the tub into the intermediate veſſel (notwithſtanding the tube of ſafety), in ſuch caſes where the heat is not kept up and urged particularly towards the end of the diſtillation, or where any negligence has taken place in the agitation which is required for the ſpeedy abſorption of the gas, or where the tubes of communication are too ſmall.

2. The greateſt addreſs and precaution are required for the proper application of the recurved tube, which on the one hand communi-

A ſhort table of ſynonimes is added at the end of this work, for the uſe of thoſe who may be unacquainted with the new nomenclature.—*P. D. C.*

The method of Bleaching, to which the preſent chapter bears reference, is deſcribed in my Chemical Dictionary, art. Bleaching.—*Note of the Tranſlator.*

cates

cates with the matrafs, and on the other with the intermediate veffel. The flighteft agitation, whether in attending the lutes, fupplying the furnace with coal, &c. is fufficient to break this tube, and likewife that which communicates to the tub. The difagreeable confequences of fuch an accident, when the apparatus is in a ftate of activity, and the difengagement of the gas muft continue to take place, are too obvious to need defcription. The fame accident may happen whenever the tube is put in its place, or taken out to clear the matrafs. To this we may add, the difengagement and frequent renewal of fo many ftoppers of cork, which are corroded by the gas and the acid during their paffage, and the adjuftment of the lutes required to cover and defend them. The care and vigilance required to maintain the feveral lutes of the whole communication muft be extreme.

3. The pneumatic tub or veffel having no cover, muft fuffer a large quantity of gas to efcape during the courfe, and particularly towards the end of the operation, which is not only attended with lofs, but renders it impoffible to remain for any length of time in the place of diftillation, without being greatly and even infufferably incommoded.

4. It is not a flight tafk to conftruct the fides or

or borders of the inverted veſſels in the pneumatic tub, to retain and concentrate the gas in a proper manner. The memoir affords no explanation of the manner of conſtructing or adjuſting theſe parts, and, conſequently, leaves a degree of uncertainty, which expoſes ſuch operators as may not be aware of the great importance of the perfect cloſure of theſe parts, to the probability of making very conſiderable miſtakes.

5. The long ſucceſſion of lixiviations and immerſions preſcribed in this memoir, which are indiſpenſible according to that proceſs, is productive of much loſs of time and inconvenience.

6. The method of compoſing the lutes, particularly thoſe which are proper for this diſtillation, not being explained, any one who is not acquainted with the means of doing this, or cannot conveniently procure them, will be much embarraſſed, more particularly if his reſidence be in the country, where the practical chemiſts of the vicinity, if any, may either be unprovided in this reſpect, or not diſpoſed to ſupply either the lute, or the inſtructions for making it.

7. Laſtly, I have found, by my own experience, that, independent of the difficulties here enumerated, the ſingle obſtacle of keeping the lutes in a proper ſtate during the whole courſe of the diſtillation,

diſtillation, together with the no leſs eſſential requiſite of preventing the danger of immerſions, are quite ſufficient to repel the efforts of the moſt zealous and obſtinate in this kind of operation.

Such are the leading impediments to which every one, whether he be a practical chemiſt or not, will find himſelf expoſed in an attempt to follow in the large way the proceſs of bleaching deſcribed in the ſecond volume of the Annales de Chimie. It was, therefore, of eſſential conſequence to diminiſh, or rather to remove theſe difficulties, without which this important art might be conſidered as of no value to the public. It will be ſeen in the account of the methods I have employed, whether I have ſucceeded in rendering the application and practice of this new proceſs much more advantageous and practicable, by perſons the leaſt acquainted with chemical manipulations.

CHAP. II.

The Methods substituted instead of those enumerated in the foregoing Chapter.

I SHALL, in the first place, describe the furnace I have made use of, which I have endeavoured to render of the greatest possible utility, without increasing the expence of fuel.

A simple cask of the proper height, or four pieces of wood framed together (see plate I, fig. 1 and 2 *), support the furnace. The hearth is disposed upon boards defended by tiles placed on a bed of clay. The walls or sides are formed of bricks, likewise connected with clay. This furnace would be equally useful and solid, and perhaps lighter, if it were lined with plaister, like those portable furnaces commonly used in Paris, which it considerably resembles in its manner of support. It is usually double, and ought in fact to be so, when the operations are intended to be made on a scale of some extent; and consequently it is divided in the middle by a partition. At the front of the furnace above

* The descriptions of the plates, by literal reference, are found at the end of this work.—N.

there

there are two openings, which may be either round or ſquare, adapted each to receive a ſquare or cylindric capſule, with a ledge, and flat or rounded at bottom. Behind, and on the ſame level as the capſule, there is a vent or pipe which conveys the heat and vapour of the charcoal, which is burned in a chafing-diſh, or upon a portable grate, beneath, or round the capſules, into a kind of reſervoir, which being diſpoſed a few inches higher than the capſules, ſerves to place a long ſquare baſon of ſheet-iron with projecting edges, which is kept filled with ſea ſalt or muriate of ſoda to the height of an inch and a half, in order that it may be dried during the diſtillation. At the two oppoſite extremities are two ſmall apertures, which are opened or cloſed as may be found expedient for the paſſage of the heat or ſmoke, and therefore operate like regiſters. In the empty ſpace, at the back part of the furnace, beneath the drying place, there is an opening in the ſide, into which troughs or boxes of ſheet-iron are put, containing the mixtures of muriate of ſoda, and mangaueſe, ready prepared before hand. In this place they are kept dry, and in readineſs to be poured into the veſſels, the evening before the diſtillation.

The opening through which the chafing-diſh is introduced, which is likewiſe on the ſide of the furnace, is not quite ſo much raiſed as the bottoms of the capſules, which, though ſupported by their rims on a level with the top of the furnace, have nevertheleſs their bottoms placed on a ſmall bar (verguillon). This opening may be ſhut during the diſtillation by a plate of iron, or any regiſter whatever which does not permit the acceſs of air from without, except at its lower part. The furnace, it may be perceived, is portable, and on that account can be placed in any part of the laboratory, as convenience or new arrangements may require.

If, in order to anſwer any particular purpoſe, or without attending to expence, the preference ſhould be given to furnaces of baked earth, I would then adviſe the uſe of ſuch as have their chimnies on the ſide, without a dome. Many furnaces of this kind have lately been conſtructed at Paris, by Laffineur, rue Mazarine. Their upper part, which is flat, and on a level with the chimney, allows the placing of capſules; and the chimney, which is at the front, renders it eaſy to take out the wood or charcoal which is put into a fire-place, provided with its aſh-hole in the ſame manner as the other furnaces which have a dome. This furnace is round, portable, leſs coſtly, and appears to me to be, in other reſpects, much more convenient than that

deſcribed

defcribed in the Annals: befides which, they may be made of any required fize.

2. Inftead of the matrafs, the intermediate veffel, and the tubes which communicate from this laft veffel to the diftilling and the pneumatic apparatus, I have fubftituted a tubulated retort, to which I have adapted a recurved neck of glafs or lead, the beak of which is placed and luted to a fmall leaden fupport in the form of a funnel; and this laft piece is adjufted to the end of a tube, of the fame metal, within the pneumatic tub, whofe lower extremity is bended to a right angle, and performs the office of the glafs tube in the apparatus of Berthollet. This tube, as well as the additional neck of the retort, may likewife be made either of pottery, ftone ware, or, which is ftill better, of porcelain.

Inftead of the retort, and its neck of glafs or lead, I have ufed, with no lefs but even with more advantage, a body or bottle tubulated at the fhoulder. Above the neck of this body or veffel is applied a pipe, which at the fame time forms the communication and the interior tube. I fhall hereafter fhew the method of difpofing this apparatus.

3. The pneumatic veffel, to which I adjuft a cover, is divided into three parts by two falfe bottoms, fixed in the veffel itfelf by means of its

its conical figure, or upon a hoop, or maſſes of wood, fixed with pins. I ſhall alſo, in the proper place, give an account of the manner of fixing and diſpoſing theſe falſe bottoms, as well as of other kinds of veſſels, not without their peculiar conveniences.

4. I have conſiderably diminiſhed the lixiviations and immerſions. In the chapter which treats of theſe ſubjects, it will be ſeen in what manner I have proceeded in ſimplifying theſe important operations.

5. I ſhall alſo deſcribe the method of compoſing two lutes, which I have found very uſeful; one made with the cake of linſeed, and the other known in chemiſtry by the name of fat lute. The latter, though more expenſive and difficult to make, appears to me to deſerve the preference.

6. By ſuppreſſing the intermediate veſſel, and by the ſubſtitution of a retort or tubulated bottle inſtead of the matraſs, together with the recurved neck or tubes of lead inſtead of the tubes of glaſs; when once the connecting part is well luted, in the manner hereafter to be deſcribed, no further trouble or inconvenience follows from the lutes, becauſe there is but one to take care of, namely, that of the beak of the neck placed on the tube which paſſes into the pneumatic veſſel.

veffel. This lute being renewed, if thought fit, at each diftillation, and a little attention being paid to apply it well, is never found to fail. The operator is, therefore, at liberty to employ his time in the lixiviations and immerfions. It will hereafter be feen in what manner I have fucceeded in removing the danger of thefe immerfions.

In this early ftage of our defcription, it is eafy to perceive how much lefs troublefome our apparatus muft prove, than that to which we have referred in the firft chapter. Neverthelefs, as it is of confequence that the inhabitants of the country, to whom my attention has been principally directed in this work, fhould be in no refpect expofed to failure in the fmalleft particular, I fhall proceed to give the moft minute accounts of the ufes and arrangement of the apparatus; and, in the firft place, I fhall treat of the lutes.

CHAP.

CHAP. III.

The Composition of Lutes.

FAT LUTE.

THE obſervations I ſhall offer on this particular lute are partly extracted from Baumé's Chemiſtry. I have thought it proper to add ſome uſeful obſervations for the ſake of beginners. Take any quantity of good grey or blue clay: I have always found fullers-earth *(argile à foulon)* excellent for the purpoſe. The clays of Gentilly and of Vanvres, near Paris, are likewiſe very good. The clay is to be dried in thin cakes, which may be ſpeedily done in an oven after the bread is drawn; the dried clay is to be pounded finely, and ſifted; a certain quantity of this clay, together with a ſufficient doſe of boiled linſeed oil, muſt then be beaten in an iron or bell-metal mortar for a long time, until the ſmalleſt lumps have diſappeared, and the whole maſs ſhall form a paſte, of an uniform colour, rather ſolid and tenacious, but, nevertheleſs, not adhering to the hands: this is called fat lute.

A large quantity of this lute may be prepared beforehand, more eſpecially when the operations

are

are to be performed in the large way, and almoſt continually. That which has been made for a twelvemonth is more pliant and better, but it muſt be kept in a cellar, in a covered earthen pot. When it has become too dry to be handled, it may be eaſily ſoftened, by firſt warming it, and afterwards beating it in the iron mortar, with as much of the boiled linſeed oil as may be found neceſſary.

The lute, which has ſerved for one diſtillation, may be uſed again, after the burned or decompoſed parts have been ſeparated: theſe parts may be known by the white or yellowiſh colour, and the dry or brittle conſiſtence which the lute aſſumes at thoſe places. Lute which is worked up again is ſo far from being worſe, that it is more flexible and tenacious. In this manner, the old fat lute, or that which has already been uſed, may be advantageouſly mixed with new lute. It is particularly eſſential that the burned portions ſhould be rejected from this mixture; if this be not done, the lute will not dry as it ought, and, ſo far from being ſoft and flexible, it will be harſh, ſhort, and continually diſpoſed to give way, by ſticking to the fingers.

When the quantity to be mixed, or kneaded up again, is very ſmall, the trouble of beating it in the mortar may be avoided, becauſe the operation

ration is performed very well, by kneading the matter with the hands. For this purpofe, a portion of the lute already kneaded in the mortar, and foaked with oil, may be taken and rolled in the veffel containing the pounded and fifted earth; the portion of earth which adheres may then be worked in; and, by a repetition of this manipulation, the mafs will fpeedily become enlarged, and muft be ftrongly compreffed, rolled out, and doubled again, until it is found that it poffeffes the requifite foftnefs and tenacity, and does not crack when doubled.

If it fhould happen that the lute fhould become too foft by excefs of oil, and clay is not at hand to correct this fault, the mafs will foon acquire firmnefs by expofing it to the open air upon parchment, or upon a plate. It muft not be laid upon paper, becaufe it is very difficult to feparate this material entirely; and if any particles fhould remain, there would be reafon to fear that, when incorporated in the mixture, they would either prevent the perfect adhefion of the lute, or would allow the paffage through that kind of void, or pore, which the fragments of paper would form. It is, moreover, to be remarked, that this lute cannot be too fmooth and uniform. It ought not to afford any perception of inequality when it is handled, or kneaded,

kneaded, nor indicate the preſence of foreign ſubſtances, ſuch as ſand, ſtraw, earthy particles, &c. which are capable of preventing the intimate connection of its parts.

I ſtrongly inſiſt on the perfection of this lute, becauſe it is the ſoul of diſtillation.

Boiled linſeed oil is thus made: two pounds of common linſeed oil being put into a ſaucepan, or proper veſſel, of copper, iron, or pottery, add three ounces of red litharge, finely powdered and ſifted; after ſtirring the whole well together, place the veſſel on the fire, heating it gradually, until the litharge is completely diſſolved. It is neceſſary to ſtir the mixture very frequently with a wooden ſpatula, until the whole ſolution, which at firſt acquires a brick-duſt colour, is completed: it is then to be removed from the fire, and, when cold, transferred into a ſtone or earthen veſſel, and kept well corked. This is the boiled linſeed oil above directed to be uſed in making the fat lute.

When this oil, which is blackiſh after boiling, is well made, it congeals in the veſſel as ſoon as it is cold. When it is required to be poured out, it may be rendered fluid by bringing it near the fire. To ſave the trouble of heating it, it may be poured, as ſoon as made, into a plate or ſhallow veſſel, or left in the veſſel uſed for boil-

ing

ing it. It is ſeldom neceſſary to heat it for the mere purpoſe of mixture; the quantities required for this purpoſe may be taken up with the fingers, or in any other manner.

It is proper to obſerve, that the veſſel in which the oil is boiled muſt be ſufficiently high, to afford a ſpace for the ſwelling of the fluid; for, as ſoon as the heat begins to act, it will riſe and overflow the veſſel, if particular attention be not paid to it. As ſoon as this proceſs begins, the veſſel muſt inſtantly be taken off the fire, and the mixture ſtrongly agitated by plunging the ſpatula in it, at the ſame time blowing ſtrongly at its ſurface with the mouth; by which means the ebullition will be checked. After this event has happened two or three times, it may with certainty be concluded, that the oil will be ſufficiently conſiſtent to form a good fat lute. By cooling, it immediately congeals, as has been remarked, to the conſiſtence of plaiſter, of a black colour, inclining to brown.

The lute made of linſeed oil cake is thus made:

The cake is firſt to be broken and pounded in an iron or bell-metal mortar, and afterwards ſifted through a ſilken ſieve; ſtarch is then to be boiled up, to the conſiſtence of ſize or glue; a ſmall piece of this, being powdered with the flour

flour of the oil-cake, is to be worked in a plate, or with the hands; more of the flour may then be added, and the kneading continued until the mafs is abfolutely without any lump, or inequality, and its confiftence has become nearly the fame as that of the fat lute; after which it is to be kept in a plate, or covered wooden bowl, in the cellar, for ufe. The fame care muft be taken with this, as with the fat lute, not to wrap it in paper, but in parchment, if thought neceffary.

This lute dries and hardens much on its outer furface, which remains uninjured at the place where it is applied; but it is decompofed more fpeedily than the fat lute, on account of its peculiar property to become hard and fhrink with a ftrong heat. In this ftate, in confequence of the action of acids, it affumes a yellow colour, and is then good for nothing: it muft be renewed.

A very good lute is likewife made with equal parts of the flour of almonds, of linfeed, and of ftarch, kneaded together. It muft be underftood, that the latter is to be boiled to the confiftence of ftarch.

To thefe different lutes we may add that which is compofed of lime and white of egg, which has the property of acquiring a confiderable degree of hardnefs.

Among all thefe lutes, that to which I have conftantly given the preference, and is always kept in fight in the prefent work, is the fat lute. The lute of white of egg and lime, retained by a cloth and a bandage, may be advantageoufly ufed as a covering to the fat lute *.

The fat lutes adhere very much to the hands, during the kneading, or working; but it is not difficult to wafh off the remains after the operation: nothing more is neceffary, than to ufe warm water and foap, or foap leys, after having previoufly wiped off the greater part with blotting paper.

* Faujas de St. Fond, in his Voyage en Ecoffe, mentions the following lute, communicated to him by the celebrated Dr. Black, of Edinburgh. That chemift confidered it as impermeable to every fpecies of gas.

This lute is compofed fimply of the pafte of almonds, in the ftate it poffeffes after extraction of the oil; it is foftened with a fmall quantity of water, in which glue has been diffolved: the glue may even be difpenfed with.—*Note of the Author*.

CHAP

CHAP. IV.

The Method of difpofing the Apparatus for Diftillation.

IT has already been remarked, that our diftillation may be performed either in a retort, or a tubulated body or bottle. There can be no difficulty in properly placing thefe veffels. The junction of the neck or tube, communicating with the pneumatic veffel, is the only object which requires particular care. The manner of joining thefe two parts, by means of lute alone, will be explained below.

As the ufe of the retort requires more attention with regard to its form, and the application of the additional part, the following details will be of ufe to prevent accidents.

When the retorts are new, and have not before been luted to any additional part, it is advifable either to rub a fmall quantity of warmed wax on the parts where the lute is to be applied, that is to fay, the neck of the retort, as well as the correfpondent part of the additional piece, or to fuffer a fmall quantity of ftarch or pafte to dry upon thofe parts; without this precaution the lute could not be eafily applied; it would flide and roll upon the glafs inftead of adhering.

Care muft afterwards be taken to fix round the neck of the retort a mafs of lute, fomewhat greater than is fuppofed to be neceffary to fill the additional part to the place where it is to be fixed, in order that by the forcing of that piece upon the neck of the retort, the lute may extend and apply itfelf more intimately. The fame attention muft alfo be paid to the mafs of lute, which is required to fecure the beak of the additional piece in its connection with the pneumatic apparatus. Thefe obfervations are of more importance, in order that the two pieces may, by this compreffion, be made to operate as if they formed one entire veffel.

To apply thefe lutes with eafe and convenience, the retort is to be held in one hand, in fuch a manner as that its belly or lower part may not touch or reft upon any thing whatever, becaufe the flighteft blow upon this very thin part will break it.

Before the lutes are applied, care muft be taken to introduce the neck of the retort into the additional piece, and mark with lute or wax upon the additional piece the place where the extremity of the retort touches it internally; and

and in like manner, on the retort itſelf, the place where the extremity of the additional piece touches its neck. By means of theſe marks it is eaſy to eſtimate the thickneſs of the maſſes of lute, by placing the two veſſels near each other in the reſpective poſitions they ought to have when fixed. Laſtly, they are united together by ſliding the recurved additional piece upon the neck of the retort, which is to be held firmly by its neck, reſting the hand on the ſurrounding part, if the retort is ſmall; or holding it by the recurved part, if it be large, or the additional piece ſhould be too long and heavy. The greateſt attention muſt be paid not to turn the parts round, during this operation, more than is abſolutely neceſſary to bring them together; and if this can be done without any turning at all, it will be ſtill better, as the lute will hold more effectually. The neck of the retort muſt be entered into the additional piece as far as it is capable of compreſſing the lute, or nearly to the marks made upon the pieces before they were put together. In this ſituation the lute, which forms a maſs round the edge of the additional piece, muſt be raiſed ſo as to cover both ſurfaces, after having firſt preſſed it as firmly as poſſible into the joint; ſmoothing it upon the two pieces, ſo as to prevent the ſmalleſt opening or crack. It is adviſable after all to ſpread a thin coating of the boiled linſeed oil over the lute, which not only renders it ſmoother and more perfect, but by the denſity it acquires from evaporation it forms a kind of varniſh or pellicle, which ſupports the lute, and prevents the fiſſures which might be formed during the actual operation. Whenever in the courſe of the work the lute ſhould appear too dry, it muſt be ſupplied with a thin coating of oil.

While the lute is thus ſpread and applied on the external part of the additional piece and the neck of the retort, the compound apparatus is to be held by the additional piece only, and the retort left to be ſupported untouched in the air, by its inſertion at the neck only.

Inſtead of luting the additional piece to the retort, ſimply at the extremity of the neck of this laſt, and at the place where the wider part of that piece touches the retort, we might apply the lute upon the whole ſurface comprehended between thoſe parts. But I have found that it is ſufficient if theſe two parts be made ſecure. A retort luted in this manner forms one ſingle and entire body with its additional neck; and with very little care and attention, the lute will ſeldom or ever have occaſion to be renewed before one or two months' ſervice.

The

The tube on which the recurved additional piece refts during the diftillation, and through which the gas is introduced into the pneumatic tube, is, as I have remarked, entirely of lead. If it be not caft, it ought to be carefully joined with ftrong folder; and, for fear this laft fhould fail, it will be prudent to cover it with a coating of yellow wax, pitch, or melted pitch.

At the beginning of my experiments, I made thefe tubes of the fame fize as thofe of the barometer; I afterwards had them of eight or nine lines (¾ inch Englifh), and did not find that the diftillation was lefs advantageous. The greater diameter even feemed to be preferable, becaufe the gas was tranfmitted with more facility, and the abforption feemed to be more effectually oppofed.

That part of the tube (if foldered as before mentioned) which paffes under the lower falfe bottom, ought to be carefully bended with a round corner, before it is coated with the wax or pitch; and in the bending it is fafer to caufe the foldered part to lie within the angle. It is likewife proper to ftop the mouth of the tube with paper, or a cork, during the time of waxing or tarring, in order to prevent any introduction of thofe fubftances into its cavity, taking care to withdraw this temporary ftopper before the apparatus is applied to aↄtual uſe. It is not abſolutely neceſſary to coat any other part of the tube, but that which is to be placed within the pneumatic apparatus, becauſe it is eaſy to ſtop any other part, out of which the gas might iſſue, with ſoft wax or lute.

The extremity of this tube, in which the recurved neck of the additional piece is to be inſerted, muſt have the form of a ſmall funnel, not only for the purpoſe of affording the moſt convenient ſupport, and the more ready adaption to the various ſizes of thoſe necks, but alſo becauſe it more readily ſupports the only kind of lute which in this work we ſuppoſe to be uſed. This lute is never deranged, if care be taken to preſs it againſt the internal ſurfaces of this ſmall funnel, and of the glaſs or lead of the additional piece, ſo as to unite them as much as poſſible, it being always underſtood that the lute is good, and poſſeſſes the properties before deſcribed in treating of that ſubſtance.

I have remarked that the uſe of the retort with its additional neck might be diſpenſed with, by ſimply uſing a body or bottle with a neck (even a wine bottle may be uſed in caſe of neceſſity, provided its bottom be either very thin, or very gradually heated). In the orifice of the neck of theſe veſſels, is to be adapted a tube of lead,

lead, properly bended, and of a due ſize. This method is in fact very advantageous and economical; but care muſt be taken to join the tube, if it be of ſheet-lead, particularly in the parts below the bottle which are liable to become heated, a ſhort time before the end of the diſtillation; to join it, I ſay, without ſolder, by fuſing the two edges together. For in proceſs of time the ſolder, though ever ſo ſtrong, yet becauſe it contains tin, is liable to exceſſive corroſion by the oxygenated muriatic acid, which, notwithſtanding its heat, is not found to attack lead in any perceptible degree.

But it may, perhaps, be more convenient to caſt ſuch a tube at one heat, as well as the additional piece in the apparatus, with the retort; unleſs, indeed, it ſhould be practicable to have it made of ſtoneware or porcelain, the latter of which is the leaſt permeable to the gas. Or we might, with more advantage, make uſe of a thick tube of common glaſs, which might be eaſily bended in a charcoal fire, and might be adapted to the tubulated bottle, as well as the leaden tube. But the danger of its breaking, and the difficulty of procuring others in caſe of need, together with the expence, have led me to reject this, as well as the tubes of pottery or porcelain.

In order that the tube adapted to the neck of

of the bottle may accurately fit, and prevent all eſcape of the oxygenated muriatic acid, it is defended by lute in ſuch a manner, that it ſhall not be thruſt into the neck of the bottle, without extruding a portion of that ſubſtance; and a border of luting muſt then be applied round the place of junction, which will effectually prevent the eſcape of any vapour which might iſſue through the firſt luting. Laſtly, the whole ſurface of this external luting is to be ſmeared with boiled linſeed oil; after which the diſtillatory apparatus may be conſidered as perfectly ſecure.

If a tube of glaſs be uſed, it may be ſo adapted by grinding with emery as to fit the neck of the glaſs body, and require no luting. The ſame might be done with a tube of porcelain, if the material were ſufficiently fine.

With regard to the other neck which I have recommended, as well in the bottle as in the retort, it ſerves not only to introduce the materials when the leaden tube is previouſly luted in, but likewiſe to admit the external air, if by chance an abſorption ſhould be perceived to take place; that is to ſay, if the water, by a diminution of the heat, which leaves a kind of vacuum, ſhould riſe from the pneumatic apparatus into the body: though even in this caſe there

there would be no reafon to fear its breaking, notwithftanding its being confiderably heated, as at the end of the operation. I have exprefsly made the trial feveral times, and always without any accident. The fluid becomes gradually heated in its paffage along the fides of the tube or neck of the diftilling apparatus, before it enters and mixes with the matter in the body itfelf; and again, if the tubulated bottle and tube be made ufe of, the water rifing through the latter and falling in the middle of that contained in the veffel, cannot directly touch the fides before it becomes mixed. But, at all events, if the fmalleft abforption be feared, it will be fufficient to raife the ftopper and return it to its place the inftant after the introduction of the atmofpheric air. Inftead of a glafs ftopper, a cork may be ufed, which muft be carefully luted round the neck, if there be any reafon to think that the vapour fhould find its way through, in confequence of the neck being not perfectly round.

With regard to the pneumatic veffel, the following is the method of placing and fixing the falfe bottoms. (See plate I. and II. and plate IX. figs. 1, 2, 3, 4, 5, and 6.)

A common wooden hoop is plained flat on the fide which is to bear the falfe bottom, and fixed

fixed within the caſk with pegs which do not paſs quite through the ſtaves. The falſe bottom, ſecured together by two dove-tails, is placed upon this hoop, and fixed there by ſimilar pegs, which penetrate part of the bottom itſelf, and by that means prevent it from either riſing or turning. The cavities between the falſe bottom and the ſides are then to be cloſed round with caulker's ſtuff (*brai ſec*), or melted pitch. It muſt be remembered, that the vertical axis with its croſs arms is to be placed beneath each falſe bottom. The arms are fixed in a mortice by means of two pins, which prevent them from vibrating or getting looſe. The leaden pipe in which the extremity of the additional neck is to be inſerted, is not to be put into its place till the firſt falſe bottom is immoveably fixed. A notch is ſuppoſed to have been cut in this bottom to admit the tube; and when it is duly placed, the vacant ſpace is to be made good, firſt with tow and then with melted pitch.

Inſtead of the wooden hoop, which affords a ſolid ſupport for the falſe bottom, it may anſwer the purpoſe very well, if cleats or blocks of wood, three inches thick, be pinned on, at different parts of the circumference; or, which is ſtill better, if the trouble be taken to fit the falſe

falſe bottom ſo well, that it may bear ſimply upon the inclination of the ſtaves, which naturally oppoſe its deſcent. This method would certainly be the quickeſt, and is not very difficult to be done.

When the falſe bottom is thus fixed, it muſt be retained in its place by pins placed at certain diſtances, and afterwards made tight by caulking.

In order that the tube may not be expoſed to vary in its poſition, a mark muſt be made on the edge of the funnel which terminates one of its extremities, by which it is eaſy to aſcertain the poſition of the bended part below, and place the ſame in the moſt favourable ſituation. It will be convenient to fix the pipe in this proper ſituation, by means of two pegs, which muſt be drawn out previous to the laſt fixing of the falſe bottoms.

When the firſt or loweſt falſe bottom is ſecured in its place, the ſecond arm of the agitator is to be faſtened to the axis, and the other falſe bottom is to be placed and made faſt in the ſame manner as the firſt.

It is particularly neceſſary to place theſe two partitions in ſuch a manner, as that the holes of communication may not be in the ſame vertical line, but as far as poſſible from each other, that is is to ſay, diametrically oppoſite. This diſpoſition is neceſſary in order that the gas may have time to concentrate in one part, before it eſcapes to the other. For the ſame reaſon, it is proper to direct the lower opening of the leaden tube to that extremity of the diameter which is oppoſite the pipe of communication from the firſt to the ſecond bottom, in caſe one diſtilling veſſel only is uſed. If two or more communicate with each pneumatic apparatus, the openings of the tubes muſt be reſpectively diſpoſed at equal diſtances, as far as poſſible from each other, and from the opening in the falſe bottom next above them.

If inſtead of falſe bottoms the preference ſhould be given to inverted tubs (*cuvettes*), the following method may be uſed to make the rims or ſides, and to fix them immoveably. (See plate IX. fig. 1, 2, 3, 4, 5, and 6.)

The rim may be made in two ways; either by ſhort ſtaves, fixed with wooden hoops as uſual, ſcarfed or hooked together at their two extremities, or elſe, by ſimply fixing a broad wooden rim, like that of a ſieve, round the bottom of this inverted veſſel, by means of ſmall wooden pins with heads.

Both theſe methods are good. The ſecond has the advantage of taking leſs room and being cheaper.

cheaper. If this method be ufed, the points of the pins muft be made a little thicker than the ftem, in order that they may be lefs difpofed to draw out of the holes bored in the bottom. With regard to the joining of the two ends of this kind of broad hoop, it may be effected very firmly by fewing them together with a flat ftrip of ofier, as is done in the better fort of chip boxes, or it may be very well managed by means of two pins with heads, which may be driven through the overlapping part, and fecured at the other fide by driving a fmall wedge into the tail of each pin. With regard to the empty fpaces or openings which may be between the rim and its bottom, they muft be ftopped with glaziers' putty (*maftic du vitrier*), which may be fmoothed with oil. This putty is of excellent fervice when the muriatic acid is ufed without potafh; but it is foon deftroyed if potafh be put into the pneumatic veffel. In this cafe the internal part of the places of junction muft be pitched or caulked, as has been already fhewn.

The method of making thefe inverted veffels with ftaves and hoops, has the advantage of being clofe, and not requiring any particular caulking.

Laftly, inftead of thefe inverted veffels, the operation

operation may be performed merely by flat boards without rims, provided, however, that the upper board be ſome inches broader on every ſide than the lower, in order that the bubbles of gas may be forced in their aſcent to ſtrike each board in ſucceſſion, and remain for a ſhort time in contact with it. The eſſential circumſtance in this arrangement will be to keep the upper part of the veſſel well cloſed, which is to be defended at the hole which admits the axis of the agitator by a central tube to retain the gas; and the partial eſcape which might take place between that axis and the covering, muſt be more effectually prevented by a cloth ſoaked in alkaline lees. This method, beſides its convenience, requires leſs care in fixing, but it renders it neceſſary to work the agitator more frequently, in order to haſten the abſorption of the gas in the water. I have determined to relate all the methods which I have ſucceſsfully practiſed, in order that thoſe who may undertake any work of this nature, may determine for themſelves, not only with regard to general motives of preference, but likewiſe the facility with which their own ſituation or circumſtances may enable them to carry the ſame into execution.

The next object is to fix theſe inverted veſſels in

in the pneumatic apparatus. This is a very ſimple operation, and conſiſts merely in fixing pieces of wood or brackets, three inches in length, under each of the two bars which connect the pieces of the bottoms of the inverted veſſels together. The bracket pieces are faſtened to the ſide of the veſſel with oak pegs, and the croſs-bars which reſt upon them are ſecured by pins of the ſame material driven above them and on each ſide, in ſuch a manner that the central perforation is in its true place, and the whole is incapable of being removed or diſturbed.

In this operation, as I have already recommended with regard to the falſe bottoms, it is adviſable to place the revolving axis in its proper ſituation, in order to aſcertain that it is not likely to be impeded in its action. It is beſt, indeed, to avoid fixing either the two inverted veſſels or the two falſe bottoms, if theſe be uſed, until the clear movement of the agitator has been aſcertained; without which precaution, there might probably be occaſion to diſplace them, either in whole, or in part, to remove the impediments which might prevent the free motion of the parts.

From the deſcription I have here given, it may be ſeen that my pneumatic veſſels have

 only

only two falſe bottoms or inverted veſſels. I think it adviſable not to uſe more, becauſe I have remarked that three of theſe veſſels requiring a greater depth, the diſtillation became much more laborious, particularly when I made uſe of the intermediate apparatus. 1. The lutes did not ſo well reſiſt the preſſure of the vapour. 2. It was not diſengaged with the ſame ſpeed, and conſequently the operation was more tedious. It is better, therefore, to uſe ſhallower veſſels, and enlarge their dimenſions in the diametral direction, as I have conſtantly found. The proportions which have appeared to me to be advantageous for a ſmall common workſhop, are 1 ½ foot in height, 32 inches in diameter below, and 36 inches diameter above, all inſide meaſure.

With regard to the kind of wood for conſtructing the veſſels, it has appeared to me to be almoſt a matter of indifference. I uſed fir, oak, and cheſtnut, without obſerving that either the one or the other were productive of any inconvenience to the quality or clearneſs of the liquor, unleſs that, at the firſt or ſecond diſtillation, the degree of force was a little altered, by ſoaking into the wood. That kind of wood may, therefore, be uſed which can the moſt readily be procured. I muſt, however, obſerve, that the large

caſks

caſks in which oil is brought from Languedoc, which are moſtly made of cheſtnut-tree, are very convenient when cut in two to form the pneumatic veſſels. They have even an advantage over the oak and fir caſks, becauſe they are cloſer in the joints, better hooped with iron and wooden hoops, and impregnated with the oil, in conſequence of which they are not ſubject to become dry, how long ſoever they may be out of uſe, provided they are kept in a cloſe place; whereas the tubs of fir wood require to be almoſt conſtantly filled with water. Oak does not contract ſo ſoon as fir.

It muſt alſo be obſerved that the white deal muſt not be uſed, becauſe it tranſmits water like a ſponge. The yellow deal is to be preferred, becauſe it undergoes leſs alteration from the fluid, no doubt on account of the reſin it contains. But if the uſe of the white deal, or any other ſpongy wood cannot be avoided, it will be proper to paint the veſſel within and without with one or two good coatings of white lead. I have had the great ſatisfaction to obſerve, that this treatment not only prevents the water from paſſing through, but likewiſe that the oxygenated muriatic acid does not attack this colour, or if it does attack it, a long courſe of time muſt be required for that purpoſe.

Melted pitch or tar likewiſe afford a good defence for ſuch wooden materials as have this defect. A mixture of yellow wax and reſin is likewiſe of excellent ſervice as a coating for the whole internal ſurface of the pneumatic veſſel, including the inverted veſſels and the agitator.

Beſides the falſe bottoms, or inverted veſſels we have deſcribed, each apparatus muſt likewiſe have its cover chamfered, to fit the circumference, with apertures to admit the tubes and the central axis; together with two others, namely, one of conſiderable ſize, to receive a funnel through which water is poured as occaſion requires, and the other ſmaller, to be opened on ſuch occaſions, in order that the air may eſcape. The cover being nailed, or rather faſtened with wooden pins, in its place, is afterwards ſecured by glueing ſlips of paper over the line where it is applied to the veſſel.

Inſtead of the wooden pneumatic veſſel, it might be more advantageous to uſe ſimilar veſſels of grit-ſtone (*grès*), rolled or caſt-lead, or cement of loriot *. Manufacturers muſt form an

* The author does not appear to ſpeak from experience in this place. It is not probable that any manufacturer would be tempted to incur the expence of ſtone veſſels; but it is nevertheleſs proper to remark, that every ſtone which could with

an eſtimate of the advantages to be derived from the expences they incur. If leaden veſſels be uſed, it will be proper to defend the ſoldered places with one or more coats of white lead, or putty, or reſin, or pitch mixed with bees-wax. I have tried theſe preſervatives againſt the deſtruction of the ſolder, and found them anſwer very well.

As it is uſeful to poſſeſs a knowledge of the height and quantity of water contained in the tub, there is a tube of glaſs fixed againſt its outer ſide, the lower end of which is bended and enters the veſſel about five or ſix inches from its bottom. This part, into which the tube is ſtuck by firm preſſure, is to be previouſly defended by lute, which is afterwards trimmed and laid ſmooth upon the ſides of the tube and the veſſel.

Laſtly, as it is eſſentially neceſſary to aſcertain, from time to time, the ſtrength of the liquor, and to draw it off upon occaſion, I have uſefully availed myſelf of a braſs cock, covered with ſeveral coatings of white lead for this pur-

with facility be wrought, contains lime or clay, or both; the former of which would no doubt be ſpeedily corroded by the liquor, on which it would alſo have a pernicious effect. It is not likely that clay would be more durable. So that on the whole there is no temptation to uſe, and many reaſons to reject, the earths.—*T.*

 poſe.

pofe. By means of this cock, it is eafy to draw off any fmall quantity of the fluid at pleafure. It has likewife the advantage of readily filling the narrow-mouthed ftone-ware, or glafs veffels, in which the liquor may be kept when there may be any to fpare, or in cafe it is thought fit to preferve a quantity always in readinefs.

When it is required to draw off the acidulated water with fpeed and in abundance, it is convenient to ufe one or more wooden tubes or fpigots, which may be opened feparately, or all at once, into appropriate veffels. But it is moft convenient that they fhould have ftoppers of cork only, becaufe thofe of wood, though covered with tow, are very apt to burft the wooden tubes by their fwelling; befides which they very feldom fit with accuracy, unlefs turned with extraordinary care.

With regard to the intermediate veffels mentioned in the Annales de Chimie, in cafe the operator is determined to ufe them, it is proper to avoid ufing ftoppers of cork to clofe the orifices, and fupport the tubes at the fame time. For this fubftance being very fpeedily acted upon by the corrofive gas, expofes the lutes and clofures to frequent derangement, as well as the tubes which pafs through them. At the beginning of my operations, I fupplied the place of of theſe ſtoppers as follows, when the necks were of a larger diameter than the tubes. I made ſtoppers of glaſs, with flanches on the ſides. Theſe were ground with emery upon the necks themſelves, and they were perforated quite through with a hole, no larger than was proper to admit the paſſage of a glaſs or leaden tube. This tube was coated with lute of ſufficient thickneſs, that it could not paſs through the hole without forming a protuberant piece, which I preſſed and ſmoothed againſt the tube as well as the orifice. Or if the ſtoppers of cork ſhould, nevertheleſs, from convenience be choſen, the necks may be covered with lute, and the ſtoppers forced in. In caſe the interval be ſmall, the parts may be heated a little, covered with virgin-wax, and then forced into the neck, and the ſmall vacuities which may remain may be filled up with the ſame wax, melted and poured out of a ſpoon. Inſtead of lute, yellow wax may alſo be uſed to fix the tube of ſafety; and the ſame operation may be performed with regard to the glaſs or leaden tube, which communicates from the tub to the intermediate veſſel. Stoppers and tubes luted in this manner, are, in ſome meaſure, fixed for ever; for when the wax is once hardened, they are in no further danger.

If the operator be ſo ſituated, that he can order the intermediate veſſels of whatever form he chuſes, it will be adviſable to have the orifices of no greater diameter than juſt to ſuffer the tubes to paſs through. No other defence will then be neceſſary, than that they ſhould be covered with lute at the time of placing them, which will render them ſufficiently firm. The rim, or border of theſe orifices, ought likewiſe to be large enough to ſupport the maſs of lute which it is proper to apply round the tube.

It may not, perhaps, be impoſſible, that ſkilful workmen, of which there are many at the glaſs-houſes, ſhould execute tubulated retorts with a recurved neck, in the form of an adopter. Such veſſels would be particularly convenient. The tubulated bottles exhibited in fig. 1 and 2, plate IX, may be ſubſtituted inſtead of retorts, with the greateſt advantage. In my lateſt operations on bleaching, I have always preferred them, becauſe more convenient, leſs coſtly, and leſs ſubject to accidents.

As it is of utility to know the method of grinding the ſtoppers here deſcribed, as well as thoſe of the tubulated retorts or bottles which may be wanted, becauſe the ſtoppers are uſually ſold in the original rough ſtate, at ſuch works as

are

are not in the vicinity of glaſs grinders, and conſequently cannot be uſed in the works we have deſcribed, I ſhall point out the method I have made uſe of. The tool which I have conſtructed for this purpoſe, plate II, fig. 1, conſiſts merely in a kind of vice or clams, in which the nob of the ſtopper may be fixed, and of which the handle being inſerted in the centre of a brace, receives and communicates the rotatory motion impreſſed upon this laſt by the hand.

As the orifices, neither of retorts, bottles, &c. nor of their ſtoppers, are accurately round, it is neceſſary, in order to reduce the firſt irregularities, that a kind of ſtopper made of iron ſhould be uſed in the firſt place before the ſtopper itſelf. Common ſand-ſtone, powdered and ſifted, may be uſed in the firſt place, and afterwards fine ſand which has been ſifted through a cloſer ſieve; or otherwiſe the ſand may be uſed firſt, and when the whole internal face of the neck has loſt its poliſh, it muſt be ground a ſhort time with fine ſifted or waſhed emery. If ſand or grit-ſtone cannot eaſily be procured, emery of different degrees of fineneſs may be made uſe of. Care muſt be taken to wet the ſtopper from time to time before it is covered with ſand, or to drop it from a ſpoon with one hand, between the ſtopper and the neck, while the

the other is employed in turning the brace. It is likewiſe neceſſary to wet the ſtopper when the grinding matter is too paſty, or the rotation takes place by ſtarts. If this be not attended to, there will be great danger of breaking either the ſtopper or the tube.

This method of grinding is expeditious enough. A quarter of an hour, or half an hour at moſt, is ſufficient for each ſtopper; but if greater expedition be required, it will be neceſſary that two perſons ſhould apply to the work, one to turn, while the other applies the grinding materials and water.

CHAP.

CHAP. V.

Preparation of the Materials.

THE knowledge of difpofing the diftilling apparatus to the greateft advantage, is not the only point in which thofe who are defirous of practifing this new method of bleaching, with the oxygenated muriatic acid, fhould be inftructed. It is very effential to prepare the materials for producing this acid, which are, as is well known, the muriate of foda, manganefe, and the fulphuric acid, or fimply the muriatic acid and manganefe. But this laft mixture is lefs convenient to be obtained, and is likely to prove expenfive, if the manufactory be not fo fituated as to obtain the acid at a low price. I fhall therefore fpeak at length of the firft mixture only, and fhall fimply obferve, with refpect to the fecond, that there is no rifk when it is ufed of breaking the diftilling veffels by the drying, and the incruftation of the materials; and alfo that the vapour of the muriatic acid, when poured out, is particularly offenfive and injurious to refpiration. It is likewife lefs eafy to procure the muriatic than the fulphuric acid, becaufe because the fabrication of this laſt in France is more widely eſtabliſhed.

The proportions of the mixture of muriatic acid and mangancſe, which I found to anſwer very well, are five ounces and a half or ſix ounces of cryſtalliſed manganeſe to one pound of the acid; that is to ſay, about two pounds and a half of manganeſe for ſeven pounds of the muriatic acid at 25°. This is the proportion ſuitable to the pneumatic veſſel, of which the capacity has been already deſcribed.

The grey muriate of ſoda * dried on an iron plate, in the part of the furnace deſcribed in a preceding chapter, and ſtirred from time to time till it appears white, is to be pounded and ſifted through a ſieve of moderate fineneſs. It is eſſential that this ſhould be done, in order that it may mix more intimately with the manganeſe, without which the ſulphuric acid could not decompoſe it with the ſame facility, and a leſs quantity of gas would be produced. The diſtilling veſſel being likewiſe ſtrongly heated towards the end of the diſtillation, there would be reaſon to fear that the coarſe ſalt might fall more ſpeedily to the

* The regulations of the ſalt duties in Great Britain are ſuch, if I miſtake not, as prevent this coarſe ſalt from being uſed in Great Britain. *T.*

the bottom than the manganeſe, and form a cruſt which would endanger its breaking.

The obſervations here made with regard to the grey muriate of ſoda, are equally applicable to the white muriate.

The manganeſe muſt alſo be pounded very fine, and ſifted through the ſame ſieve as the muriate. The good quality of this mineral ſubſtance is known from its cryſtallization, in fine brilliant needles ſlightly adherent to each other. Not that it is to be ſuppoſed that the other kinds may not produce as much oxygen, but the cryſtallized ſpecimens are to be preferred. For it is more eaſily pounded, is uſually more pure, and leſs difficult to be cleared of quartz, ſpar, &c. and likewiſe, from its cryſtallization, it preſents a larger ſurface to the action of the acid.

The white muriate of ſoda has not appeared to me to be preferable to the grey. It is true that it contains leſs impurity, but it likewiſe contains an equal quantity of water, and is dearer. The grey muriate of ſoda, on the contrary, beſides its being much cheaper, contains a certain quantity of muriate with an earthy baſis, which letting go its acid at a certain heat, ſuch as takes place towards the end of the diſtillation, permits the

the operator to uſe a leſs quantity of ſulphuric acid to decompoſe the muriate of ſoda.

The ſulphuric acid, if it be purchaſed in the rectified ſtate, ought to mark at leaſt 60° below Zero, on the Areometer for ſalts and acids of Moſſy, which is the inſtrument I ſhall always refer to in the courſe of this memoir *. As it is neceſſary to dilute the acid with its own weight of water, which then cauſes it to mark about 38 or 40°, it would be much better either to make it of this ſtrength at firſt, or to procure it without being rectified, becauſe, as Berthollet well obſerves, it would be ready for uſe, and leſs expenſive in the charges of rectification and carriage.

If the rectified ſulphuric acid be uſed, it is proper to obſerve that the acid muſt be poured into the water, and not the water upon the acid, on account of the heat and efferveſcence which is produced by the mixture. Theſe are much leſs in the firſt caſe. It is adviſable, moreover, in the mixture of theſe two liquids, to pour the acid gently, and along the ſides of the glaſs at two

* I have not been able to learn the principles upon which this inſtrument is conſtructed. It is much to be regretted that inſtruments for aſcertaining ſpecific gravities, are not conſtantly made to denote the numbers in the uſual tables where water is taken as unity.—*T.*

or

or three different times, with an interval of ſeveral minutes, taking care to turn the face on one ſide, to avoid the drops which may fly about. As the union of theſe two fluids produces, in a very ſhort time, ſuch a degree of heat, that it becomes impoſſible to touch the bottom of the veſſel with the hand, it is beſt to make the mixture in veſſels of ſtone-ware, or earth well baked, with handles, and a neck, if it be poſſible to procure them of this kind, in order that they may be lifted with eaſe and convenience, when it is required to pour the diluted acid upon the mixture of muriate and of manganeſe in the diſtilling veſſels. Common earthen veſſels may be uſed, but they are more readily penetrated by the acid, which ſplits and decompoſes them in proceſs of time.

The proportions, according to which I would adviſe the proceſs to be performed with the greateſt ſpeed and economy of time, are four pounds of muriate of ſoda, twenty ounces of cryſtallized manganeſe * (leſs of this being requiſite than of ſuch as is not cryſtallized), and forty-four ounces of ſulphuric acid at 60°, di-

* Our beſt manganeſe is brought from the duchy of Deux-ponts, from a place named Hombourg; it is uſually mixed with quartz and other matters.—*Note of the Author.*

luted

luted with three pounds and a quarter of water. I have ſometimes uſed only forty ounces of water, and did not find the operation the leſs effectual. Theſe proportions are adapted to a veſſel containing fourteen veſſels of water, each containing ſixteen Paris pints, and the materials are expoſed to diſtillation in a retort, one foot in height from the lower part of its neck, and eight inches in diameter or width in its body *. I muſt obſerve beſides, that the moſt elevated retorts are to be preferred, becauſe their roof and necks are leſs liable to be heated, and the lutes on that account are leſs liable to crack or be decompoſed. The bottles, or tubulated balloons to be uſed inſtead of the retorts, ſhould have nearly the ſame proportions. Inſtead of white glaſs, the common green bottle-glaſs may be uſed with almoſt equal advantage. For though it is leſs tranſparent, it is always ſufficiently ſo to allow the operator to ſee what paſſes within the veſſels. The only change ſuch veſſels will undergo is, that the bottom, which ought to be choſen thin, is liable to be decompoſed, or converted into the porcelain of Reau-

* This height is neceſſary, in order that one third, or one fourth part at leaſt, may remain for the ſwelling or efferveſcence of the contents.—*Note of the Author.*

mur,

mur, if the muriate of ſoda ſhould encruſt the bottom ſo as to leave no humidity between them. But this change does not render them leſs ſerviceable.

The preſent is certainly the place to ſpeak of the attempt I have made, to procure the ſulphuric acid without the intermedium of nitre, and to deſcribe the apparatus I made uſe of for that purpoſe. It conſiſted of a pitcher or pot of ſtone-ware, perforated at bottom, the neck of which communicated with two ſmall two-necked glaſs bodies connected together, and each half filled with water. Under each of theſe glaſs veſſels was lighted charcoal, to keep the water in a ſtate of evaporation, and under the earthen pot there was likewiſe fire to heat and enflame the ſulphur, which was put into the pot through the opening oppoſite the neck. This opening, which draws in the external air for the combuſtion of the ſulphur, was cloſed with a ſtopper, perforated like the nòzel of a garden-pot.

The ſulphur, thus enflamed, ſoon filled the vacant part of the glaſs veſſels with its whitiſh cloudy vapour. This vapour, meeting that of the water, combined with it, and fell in acidulous drops on the lower water, over which the vapour of the ſulphur circulating for a time, doe's alſo probably combine with it to a certain point. Another proof that this condenſed water did combine with the vapour of the ſulphur is, that the ſame vapour, received in drops beyond the ſecond glaſs veſſel by means of a recurved adopter, came out in the acid ſtate, reddening the tincture of turnſol, and effervеſcing with alkalis when concentrated. I have twice repeated this experiment with ſucceſs, and with ſcarcely any inconvenience.

I likewiſe attempted to burn ſulphur and heat water, in two ſeparate veſſels communicating with a third. The two vapours combining together in the receiving veſſel, likewiſe produced by their condenſation a fluid, which afforded the ſame indications of acidity as that of the former experiment.

When ſulphur was burned in an earthen veſſel, and its vapour communicated into an earthen jar, in which water almoſt boiling was poured, the reſults were the ſame.

It is probable that if theſe experiments were repeated more at large, with a ſuitable apparatus, a longer ſeries of glaſs veſſels, and proper furnaces, the ſucceſs would be more complete. I intend at ſome future time to reſume

ſume this proceſs, and ſhall haſten to communicate my ſucceſs to the public, if ſucceſs ſhould attend my endeavours *.

* Chaptal made a great number of experiments in the large way, for the purpoſe of diſcovering the means of acidifying ſulphur, without the expence of nitre. Upon the whole they were unſucceſsful. The manufacturer who may be diſpoſed to enter into this reſearch, is adviſed firſt to read Chaptal's paper in the Annales de Chimie, II. 86. or the full abſtract of the ſame paper in my dictionary of Chemiſtry, art. Sulphur. *T.*

CHAP. VI.

The Diſtillation.

THERE are two methods of making the oxygenated muriatic liquor, namely, with ſmell and without. Each has its advantages and inconveniences. I ſhall deſcribe both methods, begining with the liquid which has ſmell.

In the firſt place, we muſt ſuppoſe the cover of the wooden veſſel to be fixed with its pegs, and paſted round with ſtrips of paper. For this laſt purpoſe, the paſte of wheat flour is preferable to ſtarch, becauſe more tenacious.

The wooden veſſel muſt alſo be ſupplied with river water, either before or after the paſting of the ſlips of paper, which ſeldom require to be renewed. The water in the tubs, of the ſize already deſcribed, muſt riſe to the height of about ſixteen inches, or within an inch or two of the top, which may be eaſily known by the index tube on the outſide. Every inch in veſſels of that ſize anſwers nearly to ſixteen Paris pints (or very nearly the ſame number of Engliſh quarts).

Whenever this veſſel is filled with water after the

the covering is fixed and clofed, whatever may be its conftruction, it is of importance, that the fmall aperture in the covering, near that provided for the infertion of the funnel, fhould be unclofed, to permit the efcape of the internal air. This is the more effential (though the hole may be a mere gimlet hole), particularly in the conftruction with falfe bottoms; becaufe the ftream of water being conftant, when once the lower partition is filled, if the air above the fecond cannot efcape, the water, inftead of entering the veffel, will be driven back beneath the flips of paper, or along the axis of the agitator.

When the veffel is filled with water to the proper height, the retort muft be placed in its bed, upon a thicknefs of half an inch at moft, of fine fand or dry fifted afhes, and the neck of its adopter muft be adjufted in the funnel of the leaden tube. The retort is then to be fteadily fixed by pouring fand round it, to fill the vacant fpace between it and the capfule of the furnace. In this fituation the muriate and the manganefe, being previoufly mixed as equally as poffible, are to be poured through the neck into the retort, by means of a glafs, leaden, wooden, or paper funnel, perfectly dry. The goodnefs of the mixture is feen by its uniform blacknefs

 throughout,

throughout, no particles of the white muriate being in this case separately visible.

After the introduction of the muriate and the manganese into the retort, great care must be taken to clean the stopper and the neck, in order that no vapour may find its way through, when once the sulphuric acid is poured in. It is necessary, therefore, that the neck and its stopper, being previously wetted, should close the retort, as it were, hermetically. For the slightest odour would be sufficient to retard the operation, by rendering the workshop extremely inconvenient. This odour of the oxygenated muriatic acid very readily diffusing itself through the whole space of the laboratory.

When the neck of the retort or bottle, for the same observation applies to both, the beak of the adopter must be properly luted to the leaden funnel. For this purpose a piece of lute is to be rolled out in the hands, and applied round the neck of the adopter, strongly pressing it as well on the sides of the funnel as against those of the adopter, and finishing by softening each side and uniting the whole with the finger dipped in boiled linseed oil.

In the usual course of regular practice, the pneumatic vessel or vessels are filled with water, the distilling vessels duly placed and luted, and the

the mixture of muriate and manganefe introduced the evening before the day of diftillation, after having cleaned and cleared the fame veffels from their refidues *. By means of this preparatory work, there remains nothing more to be done at the beginning of the day-work, than to pour in the fulphuric acid, which during the night has had time to cool, in the earthen or ftone-ware veffels, in which it was previoufly mixed with water. Not that it would not be advantageous to pour it while yet warm, after having diminifhed its temperature, by plunging the veffel in cold water; but there would be reafon to fear, that the proper degree of heat might not eafily be adjufted, and that the diftilling veffel might be endangered by the fudden application.

It might alfo be practicable to advance the diftillation, by pouring in the fulphuric acid in the evening, immediately after the introduction of the mixture of muriate and of manganefe; but there would be reafon in this cafe to fear that the greateft part of the gas afforded by the diftillation might be loft by derangement of the lutes, if the operation were not overlooked dur-

* A fufficient number of diftilling veffels ought always to be in readinefs, to be difpofed for the diftillation of the following day. *Author.*

ing the interval. This management would not in fact be suitable to a manufactory, in which the work was not continued both day and night.

With regard to the sulphuric acid, it is to be introduced gently through a glass or leaden funnel, in order that the air which escapes may not throw up drops of the acid on the face or hands of the operator. When the acid is poured in, the neck must be stopped by turning the stopper with a slight pressure.

If the acid has been poured in warm, and the muriate is very dry, and well mixed, the sulphuric acid not more diluted than has been prescribed, and the manganese of a good quality, bubbles of air will be heard to pass into the wooden vessel, through the leaden tube, at the end of two or three minutes. If the above requisites be wanting, the escape will not take place till somewhat more than a quarter of an hour. In either case it is necessary, a few instants after the pouring of the acid, to place a chafing dish with lighted charcoal beneath the vessel which holds the retort.

About half an hour after the pouring of the acid, a considerable effervescence takes place, which sometimes swells the materials as high as the neck of the retort, if this last be too small for its charge. The bubbles of the froth are

large,

large and covered with a kind of pellicle, formed by a portion of the mixture carried up during the agitation. This intumefcence lafts about two hours, during which time the bubbles of oxygenated muriatic acid gas are moft abundantly difengaged in the water. They even fucceed with fuch rapidity, that the intervals are not diftinguifhable, and an inceffant noife is heard in the pneumatic veffel, which very often lafts three or four hours, according to the management of the fire, and the goodnefs and accurate mixture of the materials. The agitation produced by this rapid efcape is commonly fuch, that it is fcarcely neceffary to move the agitator.

The fire is not to be renewed till the expiration of two hours, even though it may have gone out in the mean time. After this, it is not to be renewed till the end of an hour and a half, and after that period at the end of an hour, and fo forth, without any perceptible increafe of its intenfity. It will be fufficient after thefe periods to keep up the fire, excepting that during the laft two hours the fire muft be maintained without fuffering the charcoal to be almoft burned away, as in the former cafes, before it is renewed. The chafing difh muft be raifed upon bricks, to bring it nearer the retort, during the laft laſt hour. I muſt obſerve, with regard to this chafing diſh, that the grate muſt not be too open, leſt the charcoal ſhould be too rapidly conſumed. After the intumeſcence of the mixture has ceaſed, the rapid eſcape of bubbles does not diminiſh for a long time, in conſequence of an effervеſcence which conſtantly proceeds. It is true that this continually diminiſhes, and towards the end of the diſtillation the bubbles which paſs into the tube appear only at intervals, notwithſtanding the matter in the retort may, by the gradual augmentation of the heat, be brought into the ſtate of ebullition. This heat is ſuch, that eight or nine hours after the commencement of the operation, the hand can ſcarcely be endured near the aperture, or the neck of the retort, or other diſtillatory veſſel, though between the fourth and ſixth hours the ſame parts are ſcarcely warm. The diſtillation of one or more retorts or bodies into a ſingle veſſel, according to the doſes before mentioned, takes uſually eleven or twelve hours, and even leſs; the time for ſtopping the diſtillation is known from the eſcape of the bubbles being very ſlow, and the noiſe leſs perceptible. This ſlight noiſe is even a mark to form a judgment of the concentration of the gas, and the degree of ſaturation of the water. In order to hear the bubbles,

bubbles, it is often neceſſary to apply the ear againſt the tub. Moreover, the adopter of the retort begins to be heated, and the lute upon its neck becomes a little ſoftened. Another indication that the proceſs is near its termination is had from the long vibrations of the water in the indicatory tube, placed on the outſide of the tub, and likewiſe in the tube of ſafety, when an intermediate veſſel is uſed.

If a proper regard be not paid to the ſigns here enumerated, and the diſtillation be not ſtopped, there will not only be a loſs of time and fuel, and a diſtillation of mere water; but the ſteam when an intermediate veſſel is uſed, will drive the water through the tube of ſafety, and itſelf immediately follow, if not inſtantly remedied by diminiſhing or removing the fire, and cooling the neck of the retort and its adopter with a wet cloth, or, which is better, by drawing the ſtopper of the retort for an inſtant.

As ſoon as the diſtillation is ſtopped, the impregnated fluid of the pneumatic veſſel is to be drawn off into tubs, or other veſſels, proper to receive the goods which are previouſly diſpoſed therein. If it be not convenient to uſe it immediately, the liquor may be left in the tub without fear of any perceptible diminution of its virtue, provided the cover and its joinings be well

well cloſed with lute and ſtrips of paper paſted on, and likewiſe that the ſpace between the axis of the agitator and the cover be ſimilarly ſecured. It may likewiſe be drawn off in ſtone-ware bottles well cloſed with corks, covered with lute at the place of their contact. In this manner the liquid may be preſerved till wanted. I have kept it for ſeveral months without its goodneſs having been impaired.

I muſt obſerve in this place, that if it be wiſhed that the liquor at the upper part of the veſſel ſhould be equal in ſtrength to that of the lower, without retarding the diſtillation (which may be uſeleſsly prolonged for upwards of twenty-four hours, by an effect of the concentration of the gas in the bottom of the veſſel, and the reſiſtance it then oppoſes to its introduction, which ſingularly contributes to increaſe the heat of the retort); I have found no better method, than that of drawing off the liquor, either into earthen pitchers or veſſels filled with merchandize ready for immerſion. I have done this after a limited time, and repeated proofs of the good quality of the fluid. At the end of eight hours diſtillation, I drew off one fourth of the contents of the veſſel; a ſecond fourth two hours afterwards; a third fourth after ten hours and a half, or eleven hours; and the reſt after twelve

twelve hours diſtillation, which formed the concluſion.

When the liquor is entirely drawn off from the veſſel, it muſt again be immediately filled with water, or at leaſt to the height of five or ſix inches above the return of the leaden tube, otherwiſe the gas, which continues to eſcape from the diſtilling veſſel and then affords no reſiſtance, might attack the pneumatic veſſel itſelf.

The fire muſt be taken from beneath the retort as ſoon as the diſtillation is finiſhed, not only to prevent the effect of the gaſeous vapours, which ſtill continue ſlowly to eſcape, from acting on the ſides of the tub, but likewiſe to diſpoſe the retorts or bodies to receive a quantity of warm water, which is to be poured in up to the neck. There is no reaſon to fear an exceſs of quantity, and the hotter the veſſels are the better. It is eſſential, however, that it be not poured in cold, for fear of breaking the glaſs. The adopter is then to be unluted from the neck of the leaden tube, if the operator chuſes; and in order that no vapour may eſcape into the workſhop, a bit of lute or a cork may be applied to the beak of the adopter. The ſand bath eaſily permits the retort to be raiſed and returned again to its place, as well as the application

plication of the lute or ſtopper to the neck of the adopter, this laſt being raiſed with one hand while the cork is put in with the other.

Nevertheleſs, as the lutes which connect the adopter with the retort are ſomewhat ſoftened towards the end of the operation, it would be more prudent to leave every thing in its place, for fear of deranging thoſe lutes. This danger is greater when the adopter is of lead, becauſe the great length of this additional piece tends to force the luting ſtill more on that account. If it be required to proceed immediately to a new diſtillation, the retort or bottle with its capſule or pan muſt be immediately taken from the furnace, and another ſubſtituted in its place ready prepared during the former diſtillation. This neceſſarily requires a double ſet of veſſels.

When the diſtilling veſſel is cold, or nearly ſo, the whole of its contents muſt be ſhaken, by holding this veſſel by the neck with one hand, and applying the other to its bottom. The ſtopper muſt then be taken out, and the veſſel ſpeedily inverted, ſhaking the reſidue to facilitate its eſcape. In this laſt ſituation the retort is to be held by its neck with one hand, and its ſide gently reſting againſt the other. The veſſels into which the water and reſidual matter of the retorts are poured, ſhould rather be of ſtone-

ware,

ware, pottery, or lead, than of wood, unleſs theſe laſt be oil veſſels, which are leſs ſubject to dry in the part above the fluid. If this circumſtance be not attended to, there will be danger of loſing great part of the contents.

It is moſt convenient to diſengage the retorts or bodies while they are ſtill warm, which continues to be the caſe the next morning after diſtillation, in conſequence of the heat of the ſand bath. If they be left to cool entirely, the ſulphate of ſoda will cryſtallize, and it will be neceſſary to diſſolve in hot water ſuch larger portions as cannot paſs through the neck. But this inconvenience is not likely to happen, unleſs the quantity of water laſt added be too ſmall, and the reſidues have been left undiſturbed for ſeveral days. The ſame obſervation is applicable to that kind of incruſtation which is formed by the muriate, if not properly pulveriſed, dried, or mixed; this cannot be ſeparated from the bottom of the retort, but by means of hot water poured at different ſucceſſive times. It is likewiſe eſſential to leave no cruſt or depoſition of muriate, or other matter, in the veſſels which are emptied, unleſs the ſame be moveable, in which caſe the riſk is leſs. But if the urgency of buſineſs ſhould then require that the ſame veſſels be uſed without entirely

tirely clearing them, it will be neceſſary to range this reſidual matter on one ſide, where it will be leſs expoſed to the heat, and will afford a greater degree of facility for the nitric acid to act upon it.

In order that the vapour which exhales from the diſtilling veſſels may not prove inconvenient, it is neceſſary to pour in a ſmall quantity of alkaline lixivium in the firſt place, which inſtantly deſtroys the ſmell. This may be done immediately after the end of the diſtillation, and the weak alkaline ſolution may ſupply the place of the water uſed for diluting the reſidues. At the inſtant of pouring this lixivial water, a ſtrong efferveſcence takes place; for which reaſon it is proper to pour it in by ſeveral ſucceſſive portions, waiting a little between each time.

The oxygenated muriatic acid obtained in this manner has a moſt keen and penetrating odour. It cannot be breathed even for a few inſtants, without the danger of a moſt obſtinate and violent cough. Its action is ſometimes ſo ſtrong that the operator will fall down ſenſeleſs, if he ſhould determine to continue his work with his noſe over the veſſels. Running of the noſe, aſthmatic affection of the breaſt, headach, tears and ſmarting of the eyes, bleeding at the noſe, the ſenſation known by the name of the

teeth

teeth ſet on edge, pains in the ſmall of the back, and even ſpitting of blood, are the ordinary inconveniences to be expected, when the pure oxygenated muriatic acid is uſed as is preſcribed in the *Annales de Chimie.* It is even impoſſible to ſupport for ſeveral ſucceſſive days an employment ſo deſtructive to the health, if the lutes be not carefully attended to, and the veſſels for immerſion of goods be not covered and placed in a ſhed, through which a ſtrong current of air paſſes. I am moreover perſuaded that there is not, perhaps, any perſon who has ſuffered ſo much as myſelf in this reſpect, on account of the earneſtneſs with which I attempted to bring this proceſs of bleaching to perfection, or rather to make it more generally uſeful. The ſtrong expectoration to which I was expoſed, agitated the ſyſtem ſo much, that I found it impoſſible to retain any food on my ſtomach, and was for forty-eight hours, without intermiſſion, not only deprived of ſleep, but continually emitting ſaliva, with acid and corroſive humours from the eyes and noſe in ſuch abundance, particularly from the eyes, that it was ſometimes five or ſix hours before I could open them to ſupport the light. My ſituation, at thoſe periods, was ſo diſagreeable, that I could not lie a moment on my back, and a very ſhort time on my ſide. The erect pcſition was leaſt painful; but I was ſoon obliged to ſit or lie down, in conſequence of the pain I felt, at every attack of the cough, in the muſcles of the back and thighs.

The difficulty, or rather the impoſſibility, of ſupporting ſuch painful exertions for any length of time, induced me to contrive a maſk of card, with glaſs eyes, which allowed me, for a certain ſpace, to work with my face over the veſſels for the immerſion of goods, to turn, preſs, and wring the pieces without fear of any ſerious inconvenience. I likewiſe occaſionally made uſe of a handkerchief, moiſtened with alkaline lixivium, which I bound round my head to defend my noſe and mouth from the effect of the odour, but theſe means were merely palliatives.

As it is of great conſequence that the operator ſhould be defended againſt ſuch accidents, or at leaſt be able to diminiſh their conſequences, it may be of ſome advantage to know, that I have had the pleaſure to experience, that the black extract of liquorice, which I chewed before I expoſed myſelf to reſpire this vapour, almoſt always produced a good effect, by diminiſhing the cough, and in ſome inſtances preſerving me from it. I therefore was particularly careful to uſe this extract, previous to expoſing myſelf

myſelf to the danger of reſpiring the gas, at the ſame time taking care not to omit the uſe of my moiſtened handkerchief, or maſk.

A ſolution of ſugar in warm or cold water, ſipped or drank ſlowly, likewiſe appeaſed the cough very much after a certain time. The warm ſolution was rather the moſt effectual. I likewiſe occaſionally ſipped or drank milk for the ſame purpoſe.

Being at length, however, worn out with ſuffering, and unable to purſue my experiments on bleaching with the requiſite convenience, I endeavoured to make the acid without ſmell, avoiding, at the ſame time, any conſiderable increaſe of expence. The following proceſs ſucceeded beſt of any that I tried. It conſiſts ſimply in adding to the quantity of water proper for each pneumatic veſſel, a quarter of a pound, at moſt, of carbonate of pot-aſh, or of ſoda, for every pound of muriate of ſoda which has been taken in the mixture of the matter for diſtillation. This quantity is ſufficient, abſolutely, to prevent the ſmell of the acid, and permit the operator to work with his face uncovered over the neutralized fluid, without riſking the ſmalleſt inconvenience. The water may be diſpoſed, for this purpoſe, in two different manners; either by previouſly diſſolving the clarified pot-aſh in the reſervoir of water appropriated to fill the veſſels, or it may be ſimply poured into the latter veſſels after it has been diſſolved, ſettled, and ſtrained by itſelf. This latter method is preferable to the other. For this purpoſe, after having diſſolved the alkali in a ſmall portion of water, it is to be poured in at different times at the commencement, and towards the end of filling the pneumatic veſſels. Theſe precautions muſt be more eſpecially attended to when the veſſel has falſe bottoms, becauſe the ſolution of pot-aſh, in that caſe, mixes leſs readily with the water. I muſt here remark, that the falſe bottoms, inſtead of being placed horizontally, muſt be ſomewhat inclined towards that ſide, in which the aperture of communication, through which the gas paſſes, is made. This inclination prevents any of the fluid from remaining in the veſſel at the time of drawing off, which might happen if there were not a decided ſlope towards the place of communication.

If it ſhould, however, be deſired to prepare a ſolution of pot-aſh proper to fill the veſſels, it muſt be diluted till it marks no more than one degree beneath zero on the areometer of Moſſy already mentioned. But this arrangement is attended with trouble, and requires more veſ-

ſels,

ſels, and conſequently more room, without producing any advantage ſuperior to that which is derived from pouring the ſtrong ſolution of pot-aſh to the water at the time of filling, according to the directions already given.

It ſometimes happens, that the laſt portions of the impregnated fluid, at the time of drawing off, have a ſlight degree of ſmell; either becauſe the agitators have not been ſufficiently attended to, or becauſe the ſalts are conſtantly diſpoſed to fall to the bottom of the veſſel. To avoid this inconvenience, about a fourth or an eighth part of the alkaline ſolution may be reſerved, not to be poured into the veſſel till half an hour before the diſtillation is ſtopped. The agitator is then to be turned, and the ſuperior liquid will be without ſmell as well as the reſt, becauſe the combination takes place inſtantly. The ſame effect will follow, if the ſolution of pot-aſh be put into the bottles or veſſels uſed to draw off the liquid: nothing more being required in this caſe, than to pour a few glaſſes of the lixivium into the receiving veſſel, which, when filled, muſt be covered up or corked. In order that the ſeparate ſolution of pot-aſh, which is poured into the pneumatic veſſel at the time of filling it with water, may not be ſubject to remain in part upon the falſe bottoms, to the prejudice of the water beneath, it is poured through wooden or leaden pipes, terminating above in a conical part or funnel, and of ſuch a length as reſpectively to communicate with the ſeveral cavities into which the veſſel itſelf is divided. By this expedient it is rendered certain, that the pot-aſh, which is required to be depoſited in the ſeveral compartments, will be ſpecially poured into each. But as the alkaline ſolution does not immediately and intimately mix with the whole of the water, but is diſpoſed to flow inſenſibly from the upper compartments to the lower, by reaſon of its weight, care muſt be taken to pour no greater quantity into the loweſt compartment than four-ſixths of the ſaline ſolution, reſerving the two other ſixths for the ſecond or firſt compartment, and forbearing to pour them in till the whole, or at leaſt the greateſt part of the water has been poured into the veſſel.

The neceſſary attention for diſtilling the acid, as well as the doſes of the materials, are the ſame whether the odour be prevented or not. The only difference conſiſts in their effects, as we ſhall hereafter ſee. The colour of both ſolutions is abſolutely the ſame. That which contains pot-aſh ſeems rather leſs limpid, particularly the firſt portions drawn off, on account

of

of the ſaline depoſition which is ſtirred up by the rapid motion of the fluid as it iſſues out. A ſimilar effect happens when the veſſel has been newly painted; in which caſe, the liquor decompoſes the paint by ſeizing the oil, and from this cauſe flows out with a ſoapy, or milky appearance.

Inſtead of pouring the pot-aſh into the veſſel, as has been deſcribed, I have very often uſed the following method. To prevent the effluvia from riſing from the veſſel in which the goods were to be immerſed, I ſimply poured my ſolution of pot-aſh into one or two pots, and afterwards ſprinkled it upon the ſurface of the liquor contained in the bleaching veſſel. This aſperſion was ſufficient to check the ſuffocating exhalation of the muriatic acid gas. I frequently uſed powdered chalk for the ſame purpoſe, and with equal effect.

I muſt obſerve, that I did not reſolve upon making this new liquor without ſmell, until after I had repeatedly aſcertained, that it is impoſſible to reſiſt, for any length of time, the difficulties which reſult from the method and the proportions deſcribed in the *Annales de Chimie*. I am even of opinion, that this method of bleaching would be renounced altogether, if the operators were literally to follow thoſe inſtructions in preference to my method, or ſome other method on ſimilar principles.

I muſt particularly remark in this place, that the agitator muſt by no means be neglected. It is neceſſary to turn it ſeveral times together at the end of every half or quarter of an hour, to favour the abſorption of the gas in the water, and to deſtroy its odour by facilitating its combination with the pot-aſh. If this be not done, in either caſe, the gas beneath the firſt falſe bottom will paſs too ſpeedily into the ſecond. This paſſage muſt be prevented as much as poſſible, until it is ſuppoſed that the water in the lower compartment is nearly ſaturated.

It muſt alſo be remarked, that when a pneumatic veſſel has once been uſed to make one of theſe liquors, it muſt not be changed in its application; for nothing more ſpeedily deſtroys the veſſel and the agitator, than alternately uſing them for both. On the contrary, when the ſame veſſel is conſtantly uſed for the ſame liquor, the action of the muriatic acid is ſcarcely, in any reſpect, perceptible.

Laſtly, I muſt remark that the ſtrength of the liquor may be increaſed at pleaſure, as Berthollet alſo obſerves, by putting leſs water into the veſſels. I have ſeveral times obtained it at ſuch a degree of concentration, as to mark be-

tween

tween ten and twelve of the areometer of Moſſy. In this ſtate its colour was evidently of a lemon yellow, a little inclining to green. This liquor contained no pot-aſh, and was deſigned for particular uſes.

CHAP.

CHAP. VII.

Concerning the alkaline Lixivium or Lees.

THE method of making the lees is not a matter of indifference, whether we attend to the ſaving of time, or of alkali. The following is that which I would, from my own experience, adviſe, and which I have ſince learned is the method uſed in Ireland. It is well known, that the white colour of the Iriſh linens is highly eſteemed. The method has always ſucceeded perfectly well with me.

Upon a kind of iron platform, compoſed of two or three concentric rings, connected together by croſs pieces of the ſame metal, as may be ſeen in the figures, 1, 2, 3, 4, Plate III, which reſts on the bottom of a boiler ſet in a furnace for the ſaving of fuel, are placed the piece-goods, thread, &c. in folds or parcels. When theſe different kinds of goods are to be immerſed together in the alkali, the piece-goods muſt always be placed at the bottom. When the boiler is thus charged, the alkaline ſolution, at the ſtrength of a degree and a half under zero of of the areometer, muſt be poured in till the maſs of goods are ſoaked, and covered to the depth of at leaſt an inch or two without preſſing them too much down. Or the alkaline ſolution may be poured in, accordingly as the goods are placed in the copper. This laſt method would be preferable, in my opinion, in all caſes where there was no reaſon to fear that the goods might lie too cloſe. To prevent theſe from riſing and floating above the ſurface of the lees, a flat cover is fitted to the boiler, which ſerves to retain the heat, and prevent any dirt from falling upon the goods.

A judgment is formed that the pieces are in a ſtate to be taken out of the fluid, when this laſt under the cover in the middle of the heap is too hot to admit of the hand being held in it, or when it ſimmers round the circumference of the boiler and throws up white bubbles, which circulate towards the centre. It is not neceſſary that the lees ſhould boil; the eſſential conditions are, that it ſhould be ſufficiently ſtrong, abundant, and hot; and that it ſhould properly penetrate the goods which are ſubmitted to its action. I have frequently thought it ſufficient, between the two immerſions in the bleaching liquor, to plunge the goods (previouſly waſhed and ſtraitened out) for a few minutes into the lixivium,

lixivium, which was very far from poſſeſſing the degree of heat above mentioned. The cloths and piece-goods, which were treated in this manner, bleached perfectly well.

In order to obtain a criterion reſpecting the time conſumed in one boiling of the lixivium, I muſt obſerve, that 3,600 French pints (or about 900 Engliſh wine gallons) of the alkaline ſolution, in a boiler ſet in the manner juſt deſcribed, will be rendered juſt boiling hot in three quarters of an hour at moſt; and if pit-coal be uſed, the quantity of fuel required, for this purpoſe, will be one-third part leſs than if wood be uſed.

When the goods are to be taken out of the copper, the cords or chains which are fixed to the exterior circle of the ſtage are to be raiſed, and hooked to the fall or rope of a crane placed on one ſide of the boiler; by turning which crane the whole of the goods are raiſed out of the copper, and after they have drained for a certain time, the maſs is conveyed and placed upon two croſs-pieces over a tub deſigned to receive the ſubſequent drainings; after which, the ſeveral pieces are preſſed or wrung, and afterwards rinſed in a ſtream, if the manufacturer poſſeſſes that convenience; or, otherwiſe, they are waſhed by means of the revolving cylinder,

linder, or other machinery. Theſe operations are to be repeated as often as the goods are taken out of the alkaline lixivium, according to the nature of the merchandize; for it may eaſily be imagined, that piece-goods, hoſiery, and thread, will require different kinds of manipulation.

As it is of eſſential conſequence to loſe as little as poſſible of the lixivium in this proceſs, it will be proper to wring or preſs the goods before they are rinſed. Piece-goods may be advantageouſly wrung by means of a fixed hook, and a handle or wooden croſs, to which a ſimilar hook is adapted that moves with the handle. See Plate IX. fig. 9. Theſe hooks being placed reſpectively at the two ſides of a trough intended to receive the waſte lees, the piece itſelf muſt be paſſed and repaſſed a number of times over the hooks, as is repreſented in the figure, until either the whole piece is thus wound up, or as much of it as can be conveniently wrung at a time.

With regard to thread, it may be preſſed or wrung with the pin; and hoſiery may be treated in the ſame manner. But it is more adviſable to wring this laſt article ſeparately by hand, unleſs the operator poſſeſſes a preſs ſuited to both the laſt-mentioned articles. By this engine the goods

goods may be cleared of the alkaline lee, with more eaſe and expedition, and with much leſs injury.

In order to economiſe the fire in the fuſion of the alkali made uſe of in new lees, as ſoon as the old has been drawn out of the boiler, which may be done by means of a ſyphon, or a cock, as may be moſt convenient, the neceſſary quantity of water may be immediately poured in with the pot-aſh or crude ſoda broken in ſmall pieces, if the purified ſalt be not uſed. In this manner the ſolution may be accompliſhed for the following day. The heat of the boiler, and its fire-place, ſuppoſing a ſmall quantity of the fire to remain, the regiſters of the furnace being ſhut, and the boiler covered up, will be ſufficient to melt the alkali in the courſe of the night. It is neceſſary to break the pot-aſh into pieces of the ſize of a nut before it is thrown in, particularly that kind which is known in France by the name of pot-aſh of York, the pieces of which are as hard as ſtones, and cannot eaſily be broken but by a mallet upon a ſtone pavement. The Spaniſh ſoda is equally hard.

The method of paſſing the goods through the heated alkaline lixivium, as here deſcribed, is particularly ſuited for works upon a large ſcale.

But

But when the manufacturer intends to confine himſelf to the bleaching of ſmall articles, ſuch as hoſe, night-caps, thread, &c. without meddling with the larger piece-goods, it will be equally advantageous to uſe a ſimple boiler, properly diſpoſed in a common fire-place, with a barrel and winch above it, as is exhibited in Plate II. fig. 4, 5, and 6. For the ſaving of fuel, this boiler may be ſet in brick-work; and like the great boiler before deſcribed, it may have a ſurrounding cavity for evaporating the old lees, which will be very uſeful if it be no more than four inches in height, and of the ſame width.

New lees, which have already been uſed for the immerſion of one piece of cloth, are not to be rejected on that account: As the ſolution loſes ſomewhat of its ſtrength on account of the matter which it extracts from the cloth, and with which it enters into combination, it muſt be reſtored by adding one-third or one-fourth of lixivium from the reſervoir, which likewiſe ſupplies that portion which was carried away in the goods, and partly recovered by draining, rinſing, or preſſure. The firſt lixivium, after two boilings, can only be poured on dyed goods, becauſe it is then loaded with extractive matter, which in a great meaſure ſaturates it, and renders it black and viſcid to ſuch a degree, that it ſometimes takes a curdled or gelatinous appearance on cooling.

The ſecond lixivium may commonly be uſed three or four times for the ſame objects, taking care to ſtrengthen it every t m with one-third or one-fourth part of new lees; after which it is to be thrown, like the others, into the reſervoir. The third and fourth lixiviums may likewiſe be uſed ſeveral times, but without adding new lees, becauſe they take up but little colouring matter. It is eaſily aſcertained by the taſte, whether they have any remaining active ſaline parts. Many perſons, particularly laundreſſes, aſcertain whether their alkaline lees has loſt its force, by rubbing it between their finger and thumb. They eſtimate its quality according to the degree of lubricity it exhibits. The quality of the two firſt lixiviums may be aſcertained to a certain point by the uſe of the areometer. When they mark three degrees under zero, they are certainly too much loaded with extractive matter, and will produce no other effect on the goods, than to give them a brown colour, to the abſolute loſs of time. It is adviſable to keep the lees at the ſame degree of force, either by addition, or by changing them altogether, for the firſt two or three immerſions of the ſame goods; diminiſhing the ſtrength, however, one-third or one-

one-fourth as ſoon as the pieces have acquired an uniform colour, which will happen, at fartheſt, at the ſecond immerſion in the bleaching liquid. When the goods have arrived at this ſtate, weaker lixiviums may be advantageouſly uſed, becauſe there remains ſcarcely any thing more of impurity to be carried off; and the ſubſequent proceſs may be conſidered merely as a brightning of the colour, by detaching the ſmall portion of impurity which may remain fixed in the texture of the thread itſelf.

The old ſaturated lees being taken out of the boilers, are to be added to the other lees of the ſame kind, ariſing from the drainings into reſervoirs appropriated for that purpoſe. This fluid is of uſe to ſoak piece-goods or thread, in order to clear them of their dreſſing, or the impurity attached to the latter during the ſpinning. With regard to the lees which are obtained by preſſure or wringing of the goods, if they be not too highly coloured, they may, as well as all the others, be added to the lees in the boiler.

As ſoon as one boiling or immerſion is made, if the lixivium be in a good ſtate, whether by the addition of freſh lees or not, the next immerſion may be immediately proceeded upon in order to take advantage of the heat, in caſe the courſe of buſineſs requires it; but for this purpoſe it is neceſſary to have another frame or platform ready prepared, with the proper quantity of goods for immerſion. But if there only remain a ſmall number of pieces which require to be plunged in the alkali, they may be thrown into the ſame bath without reſtoring it, or even heating it up again. Nothing more will be neceſſary than to cover up the boiler, and leave the goods immerſed for a ſufficient time, that they may be well penetrated with the alkaline ſolution. I have often found this manner of proceeding very convenient for piece-goods, hoſe, or thread, in ſmall quantities, without making uſe of the frame or platform.

It muſt not be overlooked, that whenever the goods are taken out, the copper muſt be examined with a ſtaff, in order to diſcover and take out any ſmall articles which might be forgotten, and would be in danger of burning if they were to remain at the bottom of the copper during the time of the ſubſequent proceſs or charge.

It muſt alſo be remarked, that it ſeldom happens that any piece which may have been ſubjected to two freſh ſucceſſive lixiviations, even though it may have been a dyed piece, will receive any advantage from a further repetition. The third immerſion, to which it might be ſubjected,

jected, with the hope of clearing it of an additional portion of impurity, will be found to produce scarcely any effect, and the liquid will take up little or no colour. It sometimes happens, that the second lixivium, even though of considerable strength, is equally ineffectual.

It is of great importance that the alkaline lees should be always as clear and limpid as possible. Their value may be estimated by observing the readiness with which they absorb the colouring matter from the goods. The tubs, in which they are kept, ought, therefore, to be made of fir; because those of oak, or chesnut, though very carefully treated with lime water, always become more or less coloured in process of time. It is true, that the colour of the lees is of, comparatively, little consequence for the first two or three immersions of dyed goods (*des pièces bises*). But this is not the case when the goods have once acquired an even colour, and require only to be brightened. It is then highly essential, for the saving of time and labour, that the lees should be as clear and limpid as possible.

The new lixivium, or lees, is good whenever it marks one degree and a half beneath zero; and I have observed, that it is not necessary it should be stronger. At a higher degree of

 strength,

ſtrength, it would ſoon become as foul or coloured as the weaker ſolution; and it is uſeleſs to conſume the alkali to no purpoſe, and communicate a dark colour to the goods, which by that means become more difficult to waſh or rinſe. Care muſt alſo be taken to rinſe them after the lees, until the water, if the operation be performed in the waſhing apparatus, flows off very clear; or, on the contrary, if the goods be expoſed to a ſtream of water, they muſt not be taken out till it has been aſcertained, in various parts of the pieces, that the water, upon wringing, comes out very clear. If this precaution be not attended to, the pieces, thus imperfectly rinſed, would be in danger of becoming yellow inſtead of white, by immerſion in the bleaching liquor; and even to acquire a very tenacious and diſagreeable ruddy tinge, either partially, if the rinſing have been partial, or totally, if it have been entirely neglected.

The activity of the fixed alkali, or pot-aſh, may be increaſed, by throwing into the boiler in which this ſalt is put for ſolution, one-third or one-fourth of its weight of well burned and very white lime, of the beſt quality; which is to be tied up in a bag or cloth. By this management, the calcareous earth is leſs capable of rendering the lees turbid; or if it ſhould eſcape, it will fall

fall to the bottom when the ſolution cools. The lime may alſo be ſeparately diſſolved or diffuſed, and the pot-aſh diſſolved in this ſolution inſtead of pure water; a method which may be preferable to the other. It is ſcarcely neceſſary to add lime to the foreign pot-aſh, moſt of them contain a certain quantity; particularly thoſe which are imported from the north of Europe, or from America.

The uſe of lime has appeared to me to produce a greater effect at the commencement, than towards the end of the bleaching. At this latter period, the different operations to which the merchandize has already been ſubjected, the cauſticity of the lees, and the ſmall grains of calcareous earth which they may contain in ſpite of every precaution, would be likely to impair the ſtrength of the goods, particularly during the operations of wringing, or the preſs. But the uſe of lime has appeared to me to be of advantage at the beginning, becauſe I have found reaſon to conclude, that the goods which are thus treated acquire a decided whiteneſs in leſs time than the others.

As it may happen, that the boilers uſed for the lixiviations may leak at the place of the rivetting, and manufacturers may find themſelves embarraſſed at a diſtance from proper workmen, I conclude, that the following method of repairing them may be acceptable. It consists in beating a certain quantity of lime, slaked by exposure to the air, and sifted with a small quantity of the white cheese called *à la pie**. The mixture is to be stirred or beaten without ceasing; and the lime must be successively added, until the mixture begins to acquire a certain consistence. The damaged part of the copper being then well cleaned and wiped, this cement is to be very firmly applied, spreading and extending it at the edges as well as possible. The cement soon hardens, and the copper may be used as before. By this means every kind of small fracture, or opening, may be easily stopped without displacing the boiler.

The same kind of lime sifted, and mixed in a similar manner with leaven, may be used with equal advantage. I have had occasion to use both these cements, and it appears that the salts exercise no action upon them; or, if they act at all, it must be in a manner scarcely perceptible.

As it may be of advantage, in certain places, instead of using pot-ash, to give the preference

* I am not acquainted with this kind of cheese; but it may easily be supposed, that experiment will point out, which of the several kinds the manufacturer may have at his disposal may be the best. I suppose a very small quantity of water is to be added.—*T.*

in France to the alkali, known by the name of *falin*, which can now be bought one-third part cheaper than formerly, before the impoſt of *peage* was ſuppreſſed; I ſhall deſcribe the manner of calcining it, in order to deprive it of the colouring matter with which it is combined. The common baker's oven may eaſily be uſed, at leaſt provisionally, inſtead of a calcining furnace. It is to be heated as if for bread, taking care only to place the wood on one ſide, in order that the part of the floor which remains free, may be well heated by the circulation of the flame. The crude ſalin is to be thrown into this vacant part by means of a ſhovel or any peel; for which purpoſe, the common peel, belonging to the oven, may anſwer very well. The alkali may, without inconvenience, be diſpoſed to the height of two or three inches over the whole clear ſurface of the floor of the oven, as far as to the diſtance of five or ſix inches from the mouth; and in order that the coal of the ſmall wood may fall thereon as little as poſſible, by ſparkling, rolling, or bounding, the largeſt pieces, or faggots, muſt be placed neareſt to the ſalt. At the commencement of the operation it is neceſſary to turn up the alkali, and renew its ſurfaces from time to time; for which purpoſe, the bended part of the poker may be very uſeful.

ful. This precaution is the more essential, because it prevents the salt from adhering to the bottom of the oven by the aqueous fusion. Some samples of this salt, at the first impression of the fire, decrepitate more or less, which arises from the muriate of soda, or sulphate of pot-ash, which is almost constantly found in every kind of incinerated vegetables; and it is sometimes the consequence of adulteration, to which that of Lorraine is more particularly subject. The oven is to be more and more heated, and the salt stirred until it ceases to emit fumes, and begins to lose its smoaky or black appearance, and becomes white throughout internally, as well as externally.

When the salt is first put into the oven, the heat ought not to be greater than for baking bread. It may be of advantage to begin the operation immediately after the bread is drawn, because the heat which is already in the oven, will render the consumption of fuel less than would otherwise have been required for the calcination.

After the salt has once become white, the fire is to be kept up as steadily as possible, without increasing it; and the greatest care must be taken, lest the salt, by too strong heat, should form itself into clots or masses. Whenever this happens,

happens, the pieces muſt be broken ſmall with the rake or poker; for the internal part of ſuch lumps, though white on the outſide, would very probably retain its dark colour. If the ſalt be imperfectly, or not at all calcined, it will afford a ſolution of a yellow-blackiſh colour, ſimilar to that of old lees highly charged with colouring matter, as is the caſe with that which has been uſed for the firſt immerſion of piece-goods, or thread. The effect of ſuch a ſolution would be very different from that of the clear and limpid ſolution, which it is abſolutely neceſſary to prepare for theſe operations.

When the ſalt appears white throughout, and ignited in certain places, when it is turned over, it will be proper to take it out of the oven. The fire is then to be gently diminiſhed till no more combuſtion remains than is convenient to give light for drawing out the ſalt. The latter is then to be drawn with an iron rake, or the bended part of the poker, to the mouth of the oven; where it may be received in troughs of ſtone, or plate-iron, or caſt-iron pots, of ſufficient ſize to contain the whole. The fire may be ſuffered to decay, until the oven ſhall have acquired the proper temperature to begin the ſecond calcination, if required; which is to be managed as before, taking care only, that the ſalt

ſalt be rather more frequently ſtirred at the commencement, becauſe the floor of the oven is always ſomewhat hotter than at the commencement of the firſt calcination. The operator muſt alſo endeavour, as much as he can, to preſerve his alkali from the ſmall pieces of charcoal which fly in ſparks from the wood, though theſe are not abſolutely of much conſequence, becauſe they ſwim on the ſurface of the water in which the calcined ſalt is afterwards diſſolved; in which ſituation they may be taken off with the ſcum which uſually riſes from this ſalt, though in leſs quantity than from the ordinary pot-aſh; which laſt ſalt is expoſed to a ſtronger heat, and is uſually mixed and calcined with calcareous earth, or other earthy matter, either to increaſe its cauſticity, or to add to its weight. The coal and the ſcum obtained in this ſolution, are not to be thrown away, but may be diſpoſed upon a cloth or ſieve over the reſervoirs containing the new lees. As theſe ſubſtances retain a ſmall quantity of ſalt, it is adviſable to pour water upon them ſeveral times; after which, they may be thrown into the oven at the next calcination.

If the calcination of this alkali be carefully managed, there will be no incruſtation upon the pavement in the oven; but if that event ſhould happen,

happen, the oven would not be lefs proper for baking bread. The only effect it will have upon this article of food, would be, that the bottoms of the loaves would be rendered more uneven. Thefe incruftations may eafily be removed by ftriking them with a hammer while they are ftill hot, or by raking them off after the oven is cold, and the alkali has began to attract the humidity of the air. In this ftate the flighteft blow will detach all the faline incruftations, which may have fixed themfelves to the pavement of the oven *.

The *falin* lofes twelve or fifteen per cent when thus converted into pot-afh, accordingly as it is more or lefs humid at the time of calcination.

The procefs here defcribed, is practicable by women as well as men. The whole was ex-

* In the fecond year of the French republic, I had occafion to direct my attention particularly to the inquiry after certain fubftances proper to afford alkali by incineration; that of the marc or ftalks of grapes appeared to me, among others, to deferve the notice of thofe who are engaged in the manufacture of pot afh. I have, therefore, inferted at the end of this work, the two memoirs which I at that time addreffed to the different committees of the national convention, which were fpecially charged to excite the zeal of the citizens towards the moft proper means of fupplying the faltpetre works with the requifite quantity of vegetable alkali.

ceedingly

ceedingly well managed by a woman, to whom I gave inſtructions. In an oven capable of baking two meaſures of flour, each weighing twenty-five Paris pounds (or twenty-ſeven pounds avoirdupois), an hundred pounds of alkali may be eaſily calcined, in three or four hours, at a ſingle heat, at no greater expence than twelve or fifteen ſols, in ſuch wood as is uſed by the bakers.

After this deſcription of calcining the impure alkali, it may not be amiſs to point out the method of obtaining the alkali from the old lees, which were formerly thrown away. For this purpoſe, I prefer the following method. Inſtead of covering with maſonry, or brick-work, that ſpace which remains above the flanch or horizontal part of the boiler for lixiviation, an additional piece is to be applied round the circumference, ſo as to form a circular cavity or external boiler. The heat which this channel undergoes from the continual action of the flame beneath, very ſpeedily evaporates the old lees, with which it is for that purpoſe filled. When the lees are thus rendered very thick by evaporation, they are conveyed to a boiler, or pot of caſt-iron, properly placed upon a calcining furnace; ſee Plate III. fig. 7, 8, 9, 10, and 11. The lees muſt not be ſuffered to become entirely

entirely dry in the circular cavity, which ſurrounds the great boiler, leſt the copper ſhould be burned or oxyded.

The old lees, when brought to a pellicle in the evaporatory apparatus firſt mentioned, are, as already obſerved, to be conveyed to the boiler of caſt-iron, where they may be heated to dryneſs without any danger of breaking the veſſel; particularly if it be made of ſoft grey iron. The reſidue muſt be ſtirred as frequently as poſſible, eſpecially towards the end of the proceſs, in order to prevent the ſalt from adhering; which, in that caſe, would render it neceſſary to dig it out with a chiſſel and mallet, unleſs there were time for it to ſoften by the humidity of the air; or the operator might chooſe to ſprinkle it with water to produce the ſame effect.

In order to take the utmoſt advantage of the heat of the evaporating apparatus, it muſt be duly ſupplied with lixivium as ſoon as that which has been evaporated to a pellicle is taken out; and in caſe there ſhould be no foul lees to evaporate, it will be proper to fill the external channel with water, to prevent the copper from burning. This hot water may be drawn off for uſe, by means of a ſyphon or caſk, when required.

The heat which is carried up the chimney may alſo be converted to an uſeful purpoſe, by placing

placing another ſmall copper over the flue, into which the old lees may be put; where they will obtain a certain degree of concentration previous to conveying them into the circular channel. In this laſt veſſel they will be ſtill farther concentrated, previous to the laſt evaporation in the iron-pot, in which they are reduced to the ſtate of pot-aſh, or *ſalin*. This ſeries of veſſels may be uſed with great profit and advantage.

CHAP.

CHAP. VIII.

Respecting the Preparation to be given to the various kinds of Goods.

THE preparation neceſſary to be given to piece goods, before they are immerſed in the oxygenated muriatic acid conſiſts, firſt, in ſoaking them about twelve hours in water to diſſolve, and clear off the dreſſing, which is uſually either common ſtarch, or thin paſte of flour. If the piece be ſoaked in a trough, it muſt be diſpoſed in very open folds, and covered with water. If the ſoaking be performed in a river, or ſtream, it muſt be expoſed to the current, after having faſtened it to a poſt fixed for this purpoſe. It would be much better to form a kind of reſervoir, defended all round by planks, as well as at the bottom, in order that the pieces might ſwim therein, without being expoſed to damage or dirt; and the water might paſs in and out by two ſluices. When the goods have ſoaked for a convenient time, they are taken out fold by fold, and then preſſed or wrung by the wring, fig. 9, Pl. IX; or, if time permits, they are ſuffered to drain on the horſe. With regard

to

to the goods ſoaked in troughs, as the water becomes very foul, it would not be amiſs to ſoak them again in freſh water till the fluid comes off very clear. In fact, there cannot be too much attention paid to clear the goods perfectly of their dreſſing. This previous treatment diſpoſes them for the ſubſequent operations, by rendering the extractive colouring matter more eaſily to be diſcharged. A fulling-mill would be very uſeful for theſe workings and rinſings.

The ſecond operation to which the piece-goods are to be ſubjected, is that of maceration; which conſiſts in ſoaking them in old lees which has been uſed, and is reſerved for this purpoſe. In the macerating troughs the pieces are likewiſe to be diſpoſed in folds; not only becauſe they occupy leſs room, but likewiſe becauſe they will be leſs preſſed together, and the maceration will be more equally effected. The quantity of lees muſt be ſufficient to cover the goods; and in order that theſe may not riſe above the fluid, they may be preſſed down by means of loaded planks, or which is ſtill better, by pieces of wood capable of being fixed ſo as not to preſs the goods, but ſimply to prevent their riſing above the ſurface. In this ſituation they muſt be left for at leaſt forty-eight hours, even in ſummer, without any fear of inconvenience. For greater certainty

certainty, however, this procefs may be managed according to the heat of the atmofphere. It is known that the maceration operates in a proper manner when the lixivium is covered with an infinity of white bubbles, and begins to emit a fetid fmell; and the piece-goods have affumed a clear yellowifh ruddy colour, inftead of the grey or dark colour they had before their immerfion. This, at leaft, is the cafe with almoft all the linens of Picardy, in confequence of the fpreading of the linens in the field, where the impurities are fo far from being detached, as when the clearing is performed in water, that they become fixed, and acquire more colour. This maceration is accelerated if the lixivium be ufed hot, as it is when rejected from the boilers. It would fcarcely be imagined how much the maceration in the old lees, when thus managed, advances the bleaching of the goods. It anfwers the fame purpofe as two immerfions in the copper; whence it is evident, how much it faves of time, labour, and expence.

Inftead of macerating the goods in the old lees, I have fometimes ufed with advantage a cold bath of lime water, or milk of lime. The grey piece-goods when deprived of their dreffing, or even without that previous preparation, were plunged in this liquid; and after having remained therein no longer than five or fix hours, they were taken out, of a ruddy yellow colour, and in a proper ſtate for the action of the lixiviums.

In order that the pieces ſet to macerate may not be too cloſely preſſed againſt the bottoms of the veſſels, bundles of white oſier twigs, or merely ſticks deprived of their bark and laid acroſs each other, may be diſpoſed upon the bottom. The ſame remark is applicable to the firſt ſoaking, required to clear them of the dreſſing. This firſt operation is performed merely to ſave the lees, for when the operator is preſſed for time, the foreign matter contained in piece-goods, that is to ſay, the ſaliva of the ſpinners, and the ſtarch of the weavers, is cleared off as perfectly as in the old lees when the goods are ſubmitted to maceration.

When the maceration is carried to the deſired point, the lees are to be drawn off, the goods taken out, and wrung or preſſed in the manner already deſcribed, and rinſed in the ſtream, or in a veſſel where it may be worked until the water, which is repeatedly poured on, comes off clear; or, in preference to this laſt proceſs, the goods may be paſſed through the fulling apparatus, if the manufacturer poſſeſſes one. See the plans and parts of this mill in the Plates IV. and V. After the fulling or rinſing, the goods are left to drain on a proper ſtage, or preſſed or

wrung

wrung to ſuch degree that they may remain only moiſt or humid. For too great a quantity of water if retained would weaken the action of the oxygenated muriatic acid.

If the operator do not poſſeſs the advantage of a fulling-ſtock, but ſimply that of a ſtream of water, the waſhing of theſe piece-goods may be haſtened, by beating them on a ſtage, level with the ſurface of water, by means of dyers' ſticks or poles, which are very well adapted for this operation.

Care muſt be taken that the troughs or other wooden veſſels in which the goods are ſoaked or macerated, be clear of every crack or ſplinter, otherwiſe there would be danger of tearing them in taking out, becauſe the fermentative proceſs occaſions them to expand and preſs againſt the ſides of the veſſels.

The obſervations we have made with reſpect to piece-goods of linen or hemp, are alſo applicable to thoſe containing cotton, taking care to proportion the time of ſoaking and maceration to the coarſeneſs or fineneſs of their texture.

Particular attention muſt be paid after the rinſing or clearing of the piece-goods, ſubſequent to the firſt ſoaking or maceration, and even after the firſt lixiviations or boilings, to rub them well with black ſoap, and afterwards to clear them out, particularly along their ſelvedges; for this part, being always cloſer than the reſt, requires to be made very ſupple, in order to open it to the action of the lees and the acid. Without this precaution it might, probably, be neceſſary after the ſeveral operations to rub them ſeparately by hand, and the bleaching would be retarded by requiring ſeveral extraordinary immerſions to prevent theſe parts from being leſs perfectly white than the reſt. If the rinſing and clearing be well performed at the commencement, the beauty and evenneſs of the white colour, and likewiſe its acceleration, will be much promoted. It is alſo highly advantageous, particularly to fine goods, that the whole piece ſhould be ſoaped and cleared after the laſt lixiviation and rinſing, till the water flows off limpid. I do not heſitate to recommend this particular manipulation, as well after the maceration as after the laſt immerſion in the lees, becauſe the whiteneſs which it affords is proportionally more brilliant and ſolid.

It is proper to add, that it will be likewiſe very uſeful to proceed with the ſame attention and care in rinſing the goods after each lixiviation. This part of the work being well performed, has more effect than is uſually ſuppoſed

on

on the ſolidity of the white colour in many piece-goods which are eſteemed in the market. Out of the various kinds I ſhall only mention thoſe called De Laval, which, after dreſſing, exhibit the moſt beautiful milk white; but which have the fault of not preſerving it when they are brought into the uſual waſh, during the courſe of wear. This fault is particularly attributed to the expoſure of the goods in the field, before they are properly cleared of the extractive matter detached by the lees.

I forgot to mention, that for numbering and marking the cloths, red ochre (*ſanguine*) may be uſed as well as lamp-black ground with oil; but the mark traced with red-ochre upon a moiſtened place, is more expeditiouſly made, and equally tenacious.

Linen, thread, hoſe, mittens, and gloves, of the ſame material, muſt undergo the preparations of ſteeping in water, and in the old lees, with the ſame care. Sewing thread, hoſe, and gloves, only require more time to arrive at the deſired point in the lees, on account of the firmneſs of their texture, which ſwells conſiderably, and acquires a degree of rigidity, that oppoſes the extraction of the impurities which this operation is deſigned to remove.

 The

The firſt thing to be done in the management of thread, is to make good the faſtenings of all the ſkeins, large or ſmall; for there are many places in which the loops are either imperfectly or not at all faſtened by the ſpinners. As thread ſwells up more than half its own bulk by wetting, it is neceſſary to looſen all the faſtenings already made, in ſuch a manner that the thread may lie very looſe in the loop; for if it were otherwiſe, there would be reaſon to fear an inequality in the bleaching. One perſon may looſen, and tie up again, all the ſkeins of one hundred pounds of thread in a day. This is work for women rather than men.

When the ſkeins are made up, a ſtring is paſſed through two or three ſkeins, according to their thickneſs; upon one end of which ſtring, a certain number of knots may be made to denote the owner of the article, whoſe name is ſuppoſed to be entered in a book oppoſite the ſame number. This aſſemblage of ſkeins is called a hank, ſee Plate II. fig. 11. An account is likewiſe taken of the weight of the thread, and the number of hanks, together with its quality, as may be neceſſary. Theſe different obſervations muſt be entered in the day-book as ſoon as the goods are received, for fear of omiſſion or miſtake. The ſame care muſt be taken to regiſter the quality,

quality, number, and weight, of the piece-goods under the name of the proprietor.

The ſtrings for tying up the hanks ought to be previouſly boiled in water, as well for the purpoſe of clearing them of the dreſſing, which the manufacturer may have applied for the purpoſe of glazing them, as to render them ſupple, and prevent their curling up when acted upon by the hot lees. The ſame ſtrings may be uſed for a long time.

The ſoaking in water requires a leſs time for thread than for piece-goods, becauſe there is no dreſſing to be diſſolved, but merely the impurity which it may have acquired during the ſpinning. The thread is to be diſpoſed in layers in the ſteeping veſſels; taking care to place the end of the ſtring of each hank upon the hank itſelf, in order that there may be no difficulty in taking them out. It is likewiſe adviſable, to place thoſe articles together which belong to the ſame perſon. Attention muſt alſo be paid, to place a kind of baſket-work, as was directed with regard to the piece-goods, in order that the lower parcels while preſſed by the upper may continue to be ſurrounded with water. Fine goods ought always to be placed the laſt; and in order that no part may riſe above the ſurface, they muſt be kept down either by a cover perforated with holes, or boards properly difpofed. Thread may likewife be foaked as well as piece-goods, by expofure to a current of water; but for this purpofe it is neceffary to pafs the hanks over poles fixed beneath the water. When the thread has remained feven or eight hours in the water, it is eafily taken out after the water is drawn off. It may then be wrung on the pin, or preffed, after having been rinfed, if convenient, in clear water.

It is then to be arranged in the fame manner, one ftratum crofs-wife over the other in the troughs for maceration, as has been directed for the fteeping; or, if the fituation and convenience of the operator permit, it may be fufpended upon fticks. But it will require more time to macerate or heat in this manner, though the goods will by that means acquire a more equal colour. When they are fuppofed to be well cleared, as we have obferved refpecting the piece-goods; they are then to be taken out, wrung or preffed, then rinfed or wafhed in clear water, and wrung a fecond time, or hung up to drain. When the thread is wrung on the pin, care muft be taken to twift the hanks three times in three different parts of their circumference, drawing it out each time with as much ftraitnefs and equality as poffible, to difpofe and

arrange

arrange the threads. This operation is likewiſe neceſſary for the perfect wringing out of the ſkeins in every part. The precaution of extending the hanks upon the pin for the due arrangement of the threads, is particularly neceſſary for double thread, which is apt to curl up by the impreſſion of the heat. If they be left in this ſituation, there would be reaſon to fear, that the bleaching would be leſs perfect in the curled parts.

Inſtead of wringing the thread on the pin, which requires much time, it would be ſtill better to clear it of the water by the beetle, or by the preſs, as by this treatment the thread would be leſs ſubject to injury.

With regard to the time of maceration for thread, it requires in general leſs time than piece-goods, by reaſon of the cloſe texture of the latter. The time, likewiſe, depends on the fineneſs or quality of the article, the temperature of the air, and the goodneſs of the lees.

With regard to ſtockings, gloves, and mittens, they require at leaſt as much time, if not more, than cloths, to be properly macerated. This muſt be managed according to the cloſeneſs of their texture, and the other relative circumſtances already mentioned.

Cotton, as well as linen thread, muſt be made into

into hanks, then ſoaked in water, and diſpoſed in croſs-layers, if troughs be uſed. Cotton imbibes water with great difficulty, and is leſs diſpoſed to adhere together than linen thread, which is not ſo ſoft and porous. It therefore always ſwims at the top of the fluid in ſpite of every precaution which can be taken, unleſs it be kept down by the methods before deſcribed. The cotton being diſpoſed in croſs-layers, and thus loaded, the troughs (which, as well as thoſe for piece-goods, ought to be ſquare, for the more convenient diſpoſition of the ſeveral articles) are to be filled with water.

On account of the difficulty of imbibing water, which depends on a certain oil, gum, or reſin, it is naturally impregnated with, cotton requires at leaſt as much ſoaking as piece-goods. Though it may not appear foul, it never fails to render the water of a darkiſh colour by its impurities. The ſoaking, likewiſe, affords a ſaving of lees, for if the cotton were to be put into the lees, without this previous treatment, the copper would hold but a ſmall quantity.

After twenty-four hours, or leſs time, of immerſion, the cotton is to be taken out, and wrung or preſſed, after rinſing in clear water, if thought neceſſary. It cannot be left to drain of itſelf, unleſs the proceſs be conducted very

leiſurely

leiſurely indeed, for it requires a long time to clear itſelf of water in this manner.

The cotton is not to be macerated. After ſteeping in water it is boiled in the lees. A good ſolution of black ſoap may on ſome occaſions be uſed inſtead of the lees; but the lees are always more effectual, and need not be made ſtronger than three quarters of a degree below zero. But the ſolution of black ſoap muſt not be neglected for thread ſoiled by the oil of the jennies or ſpinning machines; for ſome of the girls who manage that proceſs take ſo little care in greaſing the ſpindles, that the bobbins of thread are ſpotted with a thick black oil, which it is adviſable to rub, previous to lixiviation, with a good ſolution of black ſoap. If this be not done, it will be very difficult to diſcharge the ſpots: in ſpite of every care they are ſometimes viſible upon the hanks even after the bleaching. I muſt here remark, that I recommend black ſoap in preference to the white, becauſe it is more active, and does not contain thoſe ſmall ſtones, or grains, which ſometimes abound in the white ſoap, and may injure the goods, or the hands of the operator. Black ſoap, indeed, has a ſtronger ſmell, but this is diſſipated in the courſe of the ſubſequent proceſſes.

Night-

Night-caps, ſtockings, ſocks, mittens, and gloves of cotton, are to be tacked together as ſoon as received, and marked with threads paſſing through each pair; upon which a greater or leſs number of knots is to be tied, and an account taken in the day-book of every circumſtance relating to their number, quality, &c. which may be neceſſary to prevent miſtakes, or confuſion, in the ſubſequent delivery to the reſpective proprietors.

Theſe goods, when marked and tacked together, are not to be ſteeped in pure water; but, on the contrary, in a good ſolution of green or black ſoap as hot as poſſible, in order that the black and oily marks, and the impurities with which they always more or leſs abound, may be eaſily detached by rubbing them with the hands.

When the ſtockings, night-caps, &c. are taken out of the ſuds, they muſt be rinſed and cleaned in clear water, and afterwards wrung by hand. They cannot be treated in any other manner, excepting by the preſs, without danger of breaking ſome of the ſtitches.

After this treatment the ſeveral pieces may be conveyed to the troughs for immerſion in the bleaching liquor; but the work ſucceeds much better if they be previouſly boiled in the lees.

With

With regard to flax and hemp, which it may be required to bleach in the rough, it will be proper in the firſt place to give them a lixiviation, on account of the difficulty of macerating them in this ſtate, though it is not abſolutely impracticable to do it, by tying them up in ſmall ſeparate parcels. The different knots of flax, muſt, therefore, be diſpoſed on the bottom of the boiler, firſt covered with a piece of wicker work, upon which a coarſe cloth is ſpread. The different knots are to be mixed as little as poſſible, taking care to make a ſlight fold, or return, at the head of each knot. They are to be preſſed by hand the ſame as the ſtockings. For rinſing and waſhing them, it is likewiſe neceſſary that this ſhould be done by hand, holding each parcel by the head, and repeatedly plunging and moving it about in water. This, at leaſt, is the management I have thought it beſt to adopt in my trials.

It is to be remarked, that the exhauſted and unſerviceable ſolutions of the oxygenated muriatic acid may be uſed inſtead of water for the lixivium, if it be ſufficiently concentrated to mark one degree beneath zero on the areometer. I have ſometimes uſed it to advantage for cotton goods, after the laſt, and even the ſecond, boilings. This liquor was not leſs diſpoſed to become

come coloured, by diſſolving the extractive parts of thread and piece-goods, notwithſtanding the acid with which the pot-aſh appeared to be combined. It ſeems probable, that the acid is but ſlightly attached to the alkali, and may be driven off by a ſtrong heat; as, in fact, I thought I perceived in ſeparate evaporations *.

* This laſt obſervation ſeems to apply to ſuch bleaching liquor only, as may have been deprived of its ſmell by pot-aſh, and not to that in which chalk, or lime, may have been uſed. —*T.*

CHAP.

CHAP. IX.

Concerning the first Immersion.

THE first, as well as the last, immersions may be made with the acid without smell, composed with the proportion of pot-ash already pointed out; but in case this proportion should be exceeded, it must not be used but for the two first immersions. In every case these ought to be made with the acid without smell, because this liquid acts with more speed and equality. We shall presently mention the reason why, in case of a greater proportion of pot-ash, no more than the two first immersions ought to be made in this inodorous oxygenated muriatic acid.

When the immersions are to be made, if the apparatus is so placed, that the trough or back designed for that purpose is placed below the spigot of the pneumatic vessel (see Plate I. fig. 1, 2), the muriatic acid is to be drawn off to the necessary height, in order that the goods which are folded in equal folds may be covered at least two or three inches. But if the arrangement be not made in this manner, the liquor must be drawn

drawn off in pitchers, or conducted to the veſſels of immerſion by leaden or wooden tubes, provided thoſe veſſels be placed on the ground, or low enough for that purpoſe. With reſpect to what remains in the pneumatic veſſel, which is ſhewn by the degrees on the external tube or gage, after the moveable veſſel of immerſion is conveyed away on caſtors, if the bleaching liquor be ſuppoſed to be ſufficiently ſtrong, and is not immediately wanted, it may be drawn off in ſtone-ware bottles of that kind which is uſed for nitrous acid; or, if requiſite, the diſtillation may proceed to its entire termination. I muſt only remark, that when the liquor is thus partly drawn off, the diſtillation is renewed though there be no increaſe of the fire, becauſe the gas which eſcapes from the diſtilling veſſels undergoes leſs reſiſtance from the diminiſhed column of water. This is even a method, as I have already remarked, to render the liquor in the ſecond partition equal in ſtrength to that in the firſt; for otherwiſe there will always be a difference between them, which, nevertheleſs, ſpeedily diſappears when the whole of the fluid contained in the pneumatic veſſel is drawn off at once, and poured into the ſame veſſel for immerſion.

I ſhall now proceed to deſcribe the method

of

of ſubmitting piece-goods to the action of the oxygenated muriatic acid.

1. Above the veſſel for immerſion is placed the winch, or reel, uſed by dyers. The piece-goods, if there be many pieces, are ſewed together, or faſtened with ſtrings in ſuch a manner, that they form a large loop, or endleſs web, which is paſſed over the reel, ſo that by turning one part riſes as the other deſcends, and the whole length is ſubjected to the action of the acid. One workman turns the handle gently, while another ſtanding in front of the cloth, takes it by each ſelvedge, and conducts it into the liquor. Two round ſmooth ſtaves may be uſed for the ſame purpoſe. The perſon who manages the immerſion takes care to prevent the piece from folding breadthwiſe, and guides the cloth in ſuch a manner, that it may continue open and expanded as it deſcends into the veſſel.

This proceſs of turning muſt be continued for half an hour, in which time the liquor, almoſt in every caſe, has produced its whole effect in equalizing the colour. It is then taken off the reel, and left in the acid for another quarter of an hour; after which it is paſſed again over the reel, and left in the liquor till the time of taking it out, which may be done immediately, if the fluid no longer acts on the goods, and should appear still of value to operate upon other pieces. This may be immediately ascertained, either by the appearance of the fluid, of which experience will render the operator a sufficient judge, or by the proof of indigo or cochineal, pointed out in the Annals of Chemistry, of which I shall hereafter speak.

If the liquor do not retain sufficient strength for new, or dyed pieces, but enough for such as are in a more advanced state (which circumstances, and qualities, will be soon learned by experience); or if it be proper for cotton stockings, or thread only, they may immediately be immersed after having wrung the cloths by the wring over the trough, in order to save the acid they have imbibed, and bring them to a state fit for boiling. Every time the piece-goods are wrung, it is necessary to arrange them in folds on a clean table, or board, whence they may be carried and arranged in the same manner upon the stage of the lixiviating boiler.

In case there be no particular haste required to boil the goods, after it has been ascertained that the acid exerts no further action upon them; and supposing, likewise, that no other goods are at hand to be immersed in the fluid, the pieces may be disengaged from the reel, and left in in the bleaching liquor until they are wanted for the boiling. This prolonged immerſion can do no harm, and only exhauſts the acid more completely. Laſtly, when this liquor is entirely exhauſted, it is either to be thrown away, or elſe reſerved for the purpoſes hereafter to be mentioned. This firſt ſolution, it may be remarked, has uſually a ruddy yellow colour.

The veſſel, or back, for immerſion ought to be ſquare, or at leaſt, long like a bathing tub, becauſe the piece-goods are much better ſtowed in ſuch a veſſel. It ought to be ſomewhat more than five quarters long, and three quarters wide, theſe dimenſions being beſt ſuited to our piece-goods (in France). The height may be proportioned to the quantity or length of the pieces intended to be depoſited therein at the ſame time, and the maſs of fluid which muſt, conſequently, be poured in.

The mixed goods of cotton and thread, or cotton alone, being more ſuſceptible of the ſpeedy action of the acid than ſuch as are entirely of linen; and among theſe the fine being more ſpeedily affected than the coarſe, they muſt accordingly be taken out as early as experience may have ſhown to be proper for their effectual bleaching. The other goods which lie beneath, and require to remain a longer time in the fluid, will be more advantageoufly acted upon, becaufe they will be lefs preffed, and will float in a greater volume of the bleaching liquor.

Inftead of paffing the goods over the reel or winch in the manner here defcribed, which requires the time and attention of two perfons, the goods might be previoufly difpofed in a frame of one ell in breadth. This frame, fee Plate VIII. fig. 1 and 2, is provided with a number of pins with heads, over which is paffed the fame number of loops attached to the felvedge of the cloth, at the diftance of an ell afunder. By this means the pieces are fufpended in a zig-zag form, and not only occupy a fmall fpace in the veffel, but likewife become of a very even colour, in confequence of the free accefs which the acid obtains to the whole of the furface when the apparatus is plunged therein. This operation may be performed by a pulley faftened to lines which fufpend the frame. The latter may thus be eafily taken out with its charge, and be left to drain above the veffel itfelf, or any other veffel appropriated to that purpofe, in cafe the liquor fhould be found fufficiently ftrong to admit a fecond frame previoufly prepared. If this fhould not, however, be the cafe, the fluid is to be difpofed as before directed.

With

With regard to linen and hempen thread, and knit or woven goods of the ſame materials, they may be managed as follows.

Over a trough for the immerſion, ſee Plate IX. fig. 7 and 8, are placed clean poles or ſticks cleared of the bark, upon which the ſkeins of thread, ſtockings, night-caps, or mittens in pairs, are to be hung. After the acid is introduced, each hank, or pair of ſtockings, &c. is to be ſucceſſively turned, by immerſing that part into the liquor which was before upon the pole. In this manner the operator proceeds from one pole to the other, and returns ſucceſſively to thoſe goods which were firſt turned. Care muſt be taken to open them well at the time of turning, in order that they may preſent a greater ſurface to the fluid. Inſtead of turning the poles ſingly in this manner, it might be ſo managed by a band, or other mechanical contrivance, that the whole might turn together upon turning one ſingle piece of the ſet. This method would be leſs tedious and fatiguing for the workmen.

It is adviſable, that the troughs for the immerſion of threads ſhould be as nearly as poſſible of a ſquare figure, in order that they may hold a greater quantity of hanks; and the diſtance between each may be very nearly equal, for the purpoſe of exhauſting the bath with uniformity; excepting that the diſtance between the ſides of the veſſel, and the thread may be leſs conſiderable.

As the bleaching liquor is liable to loſe its gas more ſpeedily in proportion to the extent of its ſurface, it might, perhaps, be proper to have the troughs rather deep than broad, in order that the gas may be more effectually retained; and ſince it is eſſential, that the bleaching liquor ſhould act with the utmoſt poſſible equality upon the threads, inſtead of pouring it into the troughs wherein theſe are diſpoſed and arranged upon poles, it would be more advantageous to cauſe it to riſe gradually to the height of the hanks or poles; a condition which may eaſily be obtained by means of one or more tubes of lead or wood; the bended parts of which might be laid under the middle of the bottom of the trough. Theſe tubes being fixed along the internal ſides of the trough, may be furniſhed at their upper extremity with a funnel of wood, or of lead, for the reception of the fluid. After the fluid has been poured in, great care muſt be taken to keep the funnels cloſed.

Theſe are the methods which it is convenient to uſe, to ſubject threads to the action of the oxygenated muriatic acid, when the operator is in poſſeſſion of a certain quantity; but when, on the contrary, the quantity he poſſeſſes is

ſmall,

ſmall, or his operations on a ſmall ſcale, certain pieces of baſket-work, with handles, for which ſee Plate II. fig. 2 and 3, may be uſed, ſeveral of which may be placed one above the other in a round or ſquare trough of oak or fir, it being of no conſequence which of the two kinds of wood be uſed; upon each bottom a ſingle layer of hanks is to be diſpoſed, taking care that it ſhall be covered with the bleaching liquor at leaſt one inch or two in depth, and to turn them upſide down, at firſt every quarter of an hour, and afterwards every half hour; laſtly, after one hour's immerſion the thread may be taken out, if its colour be equal, and other thread put in, if the bleaching liquor continues to poſſeſs ſtrength. In a word, this proceſs is to be managed like the other already deſcribed; it muſt, however, be remarked, that the bleaching liquor may appear to poſſeſs ſome ſtrength by the teſt of cochineal or indigo, though it may not have ſufficient for the bleaching proceſs; theſe nearly exhauſted ſolutions are to be reſerved either for the kind of preparation hereafter to be preſcribed, or thrown away if no immediate uſe preſents itſelf; or otherwiſe they may be kept for uſes which I ſhall deſcribe when I ſpeak of piece-goods. The colour of the bleaching liquor which has been uſed for the firſt immerſion of linen or hempen thread or ſtockings is of a ruddy yellow, the ſame as that which has been uſed for piece-goods.

Brown or white cotton threads are to be ſteeped and turned in the oxygenated muriatic acid, in the ſame manner as thread of flax or hemp; namely, upon poles, or in baſkets; with this difference, nevertheleſs, that they require to be turned only half as often. A good half-hour is ſufficient for the firſt immerſion, after which time they are to be taken out, and other thread put in, if the bleaching liquor continues ſtrong enough for uſe, for it very ſeldom happens that the new bleaching liquor is incapable of ſerving more than once for cotton. This liquor does not undergo any remarkable change of colour.

Stockings, night-caps, gloves, mittens, and ſocks of cotton, may be very well managed with regard to the immerſion, in the ſame manner as linen or cotton thread; but as this ſort of bulky articles occupy a conſiderable ſpace, and cannot conveniently be laid on the other, it is adviſable to arrange them ſeparately in layers in the troughs, which may be of any form, either round or ſquare, though the latter form is moſt convenient, and, upon the whole, to be preferred. Theſe articles are to be diſpoſed in layers upon platforms of clear oſier work, provided, as has

already

already been observed, with four handles, upon which the other platforms are to be placed. No more than three can be put into one trough. As the articles placed upon the uppermost platform might rise to the surface, which would expose them to an inequality of colour, another platform, or piece of basket-work, with a rim, may be placed above them, which must be so managed as to press the goods in a small degree, and prevent their rising. Two or three ranges of night-caps, stockings, &c. are sufficient upon each platform.

It is easily known when the cotton stockings, or night-caps, have remained a sufficient time in the first bleaching liquor. Nothing more is necessary for this purpose than to hold them up to the light; in which position they ought not to shew those opaque spots, which are of a more or less ruddy colour, according to the nature of the goods, or at least very few of those spots should appear.

Cotton stockings, with clocks, are more difficult to bleach in that part, and must be carefully pulled open every time they are immersed in the liquor, because they are very subject to shrink up. It is advantageous to turn them inside-out previous to the second immersion.

The present remark with respect to cotton stockings

ſtockings is ſtill more ſtrongly applicable to ribbed thread ſtockings. The fingers of gloves are likewiſe more difficult to bleach at their extremities, becauſe the texture is cloſeſt at that part. It is even prudent to turn ribbed ſtockings inſide-out ſeveral times during the courſe of the immerſion, for which reaſon it will be moſt convenient to place them always near the top of the veſſel. Common cotton ſtockings, and other goods, may remain in the fluid without being turned during their immerſion, becauſe they are more looſe and ſpongy. They may be left in the liquor about half an hour. Cotton manufactured into goods is more difficult to be penetrated than the ſimple thread. By means of cords paſſing through the handles of the lower platform, upon which all the others reſt, the whole ſyſtem may be very eaſily raiſed by a pulley. In this ſituation they muſt be left to drain above the trough; after which, the pieces are to be preſſed ſeparately by the hand, or all at once by mechanical means, if the operator be provided with an apparatus.

If the acid be ſtill good, other ſtockings are to be immerſed in it, either in their firſt ſtate, or in different ſtages of the proceſs; the raw articles muſt be immerſed in leſs quantity than thoſe which are partly bleached: if the liquor

be

be nearly exhaufted, it is to be referved purfuant to the recommendation already given.

The obfervation which we have made with regard to night-caps, ftockings, &c. made of cotton, in which the greater or lefs effect may be feen by holding them up to the light, is alfo applicable to gloves and ftockings of linen thread: but, as it has already been remarked, thefe goods are much more difficult to bring to an equal colou ; for however loofe the texture may be, the linen thread always fwells confiderably, fo as to render ftockings ftiff and inflexible. The texture in this fituation is fo difficult to be penetrated, that the bleaching is as it were entirely fuperficial. It is better, therefore, when ftockings are required to be well bleached, that the thread fhould have been cleanfed at leaft before the knitting or weaving, by which means it becomes more difpofed to open and imbibe the acid. Stockings, of linen thread entirely in the raw ftate, without having been cleared of their firft impurities, are very unpleafant for the bleaching liquor to operate upon, and ftill more when they are ribbed or have clocks. Thefe goods are liable to a very unequal colour.

In general, all kinds of looped or ftocking work of flax or hemp, muft be fuffered to remain

main in the liquor at leaſt half as much longer than other goods; that is to ſay, from two to three hours, for the acid does not penetrate them and give them an equal colour, unleſs it be ſuffered to operate for a conſiderable time. They muſt not even be lixiviated until their colour is nearly equal. If it ſhould happen that they do not acquire an uniform colour during the firſt immerſion, they muſt have a ſecond, which muſt, in both caſes, be of conſiderable ſtrength, and in which they muſt be kept a ſufficient time to undergo the effect without the intermediate action of the lees. This obſervation is equally applicable to all other piece-goods of linen, or ſtockings of cotton.

With regard to knots of flax or hemp, they are to be bleached in the ſame manner as the ſtockings and night-caps, by diſpoſing them as much as poſſible in thin ſtrata; becauſe filaments are naturally much diſpoſed to become entangled, and form a cloſe maſs. Knots of flax are bleached very ſpeedily, that is to ſay, by one or two immerſions leſs than are required for thread of middling fineneſs. It muſt, nevertheleſs, be obſerved, that they muſt not be bleached until after having been beaten and combed, becauſe they muſt always be ſoaped after the bleaching, on account of their adher-

adhering together while drying, a circumſtance which can hardly be prevented. If this be not attended to, there will be conſiderable loſs. Knots of flax bleached in this manner, and afterwards combed, appear to the eye as beautiful and ſhining as white ſilk.

It muſt be remarked, that if the ſtoop of flax or hemp obtained from this bleaching, or bleached ſeparately, be cut, in caſe the ſtaple be too long, and afterwards carded, it has a ſingular reſemblance to the cotton of Siam, which is very plentiful in the market, and known to have the ſhorteſt ſtaple. When it is well carded, no difference can be perceived between the two articles; neither is it poſſible to diſtinguiſh them in ſpinning. I have had an opportunity to weave ſome of this thread at the end of a web of cloth, where it might have been taken for real cotton. I have likewiſe had an opportunity of uſing it in candle wicks, in which there was no perceptible difference between it and cotton, either with regard to the colour, or clearneſs of the light. It will, no doubt, be a very intereſting object to aſcertain all the advantages which the commercial world may derive from this application of the oxygenated muriatic acid.

CHAP.

CHAP. X.

Inſtructions with regard to the Quantity of Lixiviations and Immerſions.

THE number of immerſions for hempen or linen goods is commonly three for fine goods, ſuch as hollands, cloths, lawns, &c. &c. five for common cloths, and ſeven for the coarſeſt. It may alſo happen, that an immerſion extraordinary may be required for each of theſe kind of goods, according to the accidents they have met with, the greater or leſs degree of cloſeneſs in their texture, and the dark colour of threads here and there paſſing through the cloths, particularly in thoſe known at Laval under the name of toiles brindellées. This name is given to them on account of threads paſſing through them, which are ſaid to be dyed by the manufacturer for the expreſs purpoſe of rendering it heavier, and on this account more advantageous in the ſale. The dark threads of theſe cloths can never be bleached by the common method; whence a judgment may be formed of the advantage of the new method in bringing theſe goods

goods into the market, which, though fine and equal in beauty and general whitenefs to thofe of Flanders, Ireland, and Silefia, are neverthelefs greatly depreffed in price, on account of the fingularity in the colour, which renders them at leaft twenty per cent lefs valuable.

From the number of immerfions here prefcribed, it will follow that the lixiviations may be reduced to two for fine cloths, four for middling cloths, and fix for the coarfeft kind, fuppofing the moft perfect white to be required; for if a commoner colour fhould be thought fufficient, one lixiviation, and one immerfion, may be deducted from each kind of goods; whence it follows, that for a middling white no more will be required than to give one or two immerfions to the fine goods, two or three for thofe of medium finenefs, and three or four for goods of the moft inferior quality.

With regard to piece-goods of cotton, the coarfeft will not require more than four immerfions, and three lixiviations. For fuch as confift of linen and cotton mixed, no regard muft be paid to the cotton, but to the thread, which always, during the procefs, remains behind in its degree of perfection. Neverthelefs thefe are bleached more fpeedily than if they were entirely of linen, becaufe the cotton, which is intermixed,

termixed, renders the goods more penetrable by the acid. In general no more than five immersions and lixiviations are required for the coarsest goods of this kind. The same advantage of speed is also obtained in other open-worked goods, which admit the acid more readily into their texture.

Linen and hempen threads are affected nearly in the same manner as piece-goods; that is to say, the fine thread requires no more the three immersions, and two lixiviations, the middling four or five, and the double or sewing thread, or threads of coarse quality, six or seven immersions; whence it follows, that three or four lixiviations are sufficient for fine thread, five or six for close coarse thread, and six and a half for sewing thread of the same quality. The latter threads requiring more care and attention, are likewise more difficultly penetrated by the acid.

Gloves and stockings of hemp or linen follow nearly the same proportions, with the addition of half a lixiviation and one immersion more, according to their quality, the closeness of their texture, and the inequality of the thread. Ribbed stockings, or such as have woven clocks, will, likewise, in some cases require an additional immersion. The same proportion is to be observed in these goods when they are mixed, as was ſhewn with regard to mixed cotton goods, excepting that an extraordinary immerſion, or half immerſion, is given on account of the linen thread, which ſwells up by moiſture, and always becomes white ſomewhat more ſlowly. But, on the whole, ſingle threads bleach more quickly than piece-goods, becauſe the threads are more diſengaged and ſeparate from each other, and being leſs compreſſed admit the fluid into contact with greater facility, with the exception only of dyed or ſewing thread. But this facility in the bleaching is fully counterbalanced by the care which thread requires to prevent its becoming entangled or broken.

Three immerſions are ſufficient to bleach the coarſeſt cotton thread, ſuch as that which is intended for cotton wicks; and accordingly no more than two are required for common threads with the appropriate lixiviations, it being always underſtood that the fineſt white colour is here meant. It is of no conſequence whether the cotton be of a dark colour, or inclined to white: the latter, which is naturally more foul or impure, might be expected to bleach more ſpeedily, but it frequently acquires the proper degree of whiteneſs more ſlowly than the other.

Gloves, mittens, ſocks, night-caps, and ſtockings, of cotton, require no more than three immerſions, and ſometimes two are ſufficient, according to the quality and cloſeneſs of their texture. Hence it may be obſerved, that the number of lixiviations cannot exceed two for the moſt common goods, and it is, therefore, eaſy to regulate the proceſs for an imperfect white. This colour, however, is ſeldom required on cotton.

I muſt obſerve, that by the words half lixiviation, I underſtand that the lees poſſeſs no greater ſtrength than one degree at moſt for thread and piece-goods, and half a degree, or three-fourths, for cotton, if the lees be new; but otherwiſe the operator may uſe ſuch as have already been applied, and have not been reſtored to their original ſtrength. When the ſame term is applied to immerſions, I mean to ſpeak of the bleaching liquor, diluted with one-fourth of its weight of water, or ſuch as has already been uſed for the firſt white, and ſtill retains ſtrength enough for the immerſion of pieces already advanced in their bleaching.

When the muriatic acid without ſmell is well made, the operator ſees with pleaſure that one quarter of an hour after the immerſion of thread, a white, and, as it were, ſoapy lather comes up to the top. This is a good ſign, for it very ſeldom

ſeldom happens, that pieces immerſed in a bleaching liquor which produces ſuch an effect, do not obtain an even colour. I muſt, moreover, remark, that it is not neceſſary to dry the goods before their immerſion in the lees, or the bleaching liquor. It is ſufficient that they be well wrung, or cleared of their water to ſuch a degree, as only to remain moiſt. We might even plunge them into the bleaching liquor immediately after their rinſing, or wringing out of the lees, if it were not that this management diminiſhes its ſtrength in ſome degree.

On the ſame principle we may plunge the goods, when taken out of the bleaching liquor, into the lees without rinſing, but merely after ſtrong preſſure, though the rinſing appears to deſerve the preference. To ſave time and trouble, however, I would adviſe the operator to omit the rinſing when he is deſirous of haſtening his work; the only riſk which this omiſſion affords, is that of weakening or neutralizing the lixivium to a greater degree, which by this means will not ſerve for ſo many boilings. It is proper alſo to remark, that if a lixivium thus neutralized, but not loaded, with colouring matter (which may be productive of deception if the ſtrength be not aſcertained by the taſte) be uſed, the goods will come out dyed of a

 nankeen

nankeen colour, and the operator will be aſtoniſhed that they do not bleach though ſteeped in a new and ſtrong bleaching liquor. This laſt, on the contrary, ſerves only to deepen the nankeen tinge; but, as I have before remarked, this accident does not happen, excepting when the lixivium is entirely exhauſted, and neutralized by frequent immerſions of goods therein. This effect does not uſually happen, until after the lees having been uſed five or ſix times without being renewed. The remedy for this accident will be given hereafter.

I ſhall conclude the preſent chapter by obſerving how uſeful it is to rinſe, and cleanſe the goods as ſoon as they have undergone their lixiviation, that is to ſay, a few minutes after taking them out; they are at that time more open in their texture, and more diſpoſed to part with the impurities which the lixivium may have detached.

In the rinſing of threads, they muſt not be held by the ſtring of the hanks, but, on the contrary, the hand muſt be paſſed through all the ſkains, and thus held, they muſt be ſtirred round in the water. By this treatment the thread is better cleanſed, becauſe it remains leſs entangled, and more open. If the operator have the advantage of a river, or ſtream, the

ſhorteſt

ſhorteſt method will be to loop them all on a pole, and hold them ſuſpended in the water. The poles are to be fixed in an oſier baſket, in order that ſuch hanks as may be accidentally detached, during the act of turning, or placing them, may not be carried away by the ſtream. This is much more expeditious, and leſs embarraſſing.

CHAP. XI.

An Account of the Quantity of Linen, and Cotton Thread, bleached at each Immerſion, and the Colours acquired by thoſe Subſtances.

THE quantity of pounds of linen, or hempen thread, which may be paſſed into a bath compoſed of the whole contents of a pneumatic veſſel of the acid without ſmell, obtained according to the doſes and proportions before preſcribed, may be eſtimated at ſixty or ſixty-two pounds for the firſt immerſion, and eighty for the ſecond and following immerſions. In order to avail himſelf of this datum, for piece-goods which are to be plunged in the liquor, the operator muſt take care to weigh them beforehand, previous to ſteeping them to clear off their dreſſing. This quantity is alſo ſuſceptible of variation, according to the quality of the thread. Thread of middling quality is here meant.

The quantity of cotton which may be paſſed through a ſimilar doſe of the fluid, is from eighty to ninety pounds, of middling quality, for the firſt

first immersion, and one hundred for the second. According to this rate the operations must be regulated for other objects, such as stockings, night-caps, gloves, &c.

It is more adviseable to diminish than increase this quantity of goods, to have them more perfect, more equal, and of a better white. The succeeding immersions will produce a greater effect upon threads thus treated.

The bleaching liquor which has been used for cotton becomes slightly charged with colouring matter, and at the first immersions acquires a pale amber colour. The latter immersions do not change it, but leave it clear and limpid. The same observation is applicable to both the oxygenated acids; that is to say, the acid with smell, and that which is without.

As it is of essential consequence to be aware of certain events, or facts, by which the progress of the bleaching may be ascertained, I shall here point out the gradations of colour, which the pieces assume after each immersion in the oxygenated muriatic acid without smell, made according to the proportions here described. The first immersion gives the thread, or piece goods, a reddish colour, slightly inclining to yellow; the second, a colour inclining to ruddy yellow;

the third, a whitiſh yellow; the fourth, a white, ſlightly inclining to a ruddy tinge; and by the fifth and ſixth, the white becomes clearer and clearer. Theſe are very nearly the ſhades which are aſſumed by coarſe goods, for the fine goods frequently paſs to the ſecond or third gradation by one ſingle immerſion.

When the liquor is ſtrongly concentrated in pot-aſh, ſuch as that which is denoted in the annals of chemiſtry by the name of Javelle, the goods immediately, and without previous lixiviation, aſſume the third colour; but I have obſerved, that it is difficult to bleach them further without uſing the ſulphuric acid, to remove the lees with which they are loaded. It muſt, moreover, be remarked, that in order to obtain this tone of colour, it is ſufficient that the lixivium be diluted with water, ſo as to mark two or three degrees only on the aerometer inſtead of eighteen or twenty, which it may mark after it is prepared by diſtillation.

There are ſome who do not approve the colour which the thread acquires after the firſt immerſion, but it may immediately be reduced by ſteeping the goods in cold or hot lees. The latter produes its effect more ſpeedily; and after ſubſequent rinſing and drying, the goods retain a grey white colour, more or lefs deep according to the fhade it has received. Many venders prefer this grey, or reduced colour, on account of its preferable fale in certain markets.

With regard to the bright and perfect white, there are very few perfons in the provinces who care for it, or appear to give it an exclufive preference. Two reafons may be given for this: firft, becaufe a prejudice is unfortunately eftablifhed againft the fpeed with which the new invented method of bleaching operates: and fecondly, the confumer is conftantly perfuaded, whether the bleaching may have been performed in this manner, or in the field, that when the goods have attained an extreme degree of whitenefs, they cannot be as durable as fuch as are lefs white. It is thought to be rotten, or burnt, and this opinion leads to a preference in favour of fuch linens, and even cottons, which preferve after bleaching a folid fhade of grey, or dulnefs in the white.

From a prejudice of the fame kind it is, that, in many countries, the women, particularly the peafants, prefer their linen, whether for clothing or houfehold ufe, fimply cleared without bleaching. The orders of proprietors, or purchafers, muft therefore be attended to, and the number

number of immerſions and lixiviations regulated accordingly.

It may be conſidered as a rule, that when the goods no longer communicate a perceptible colour to new lees, they are entirely finiſhed, and conſequently, that every ſubſequent lixiviation, or immerſion, will be attended with abſolute loſs, unleſs the immerſion is neceſſary to clear off the laſt lees, on the ſuppoſition that ſimple rinſing in a large quantity of water may not be ſufficient.

I muſt, nevertheleſs, remark, that thread bleached by the oxygenated muriatic acid, may be uſed by the ſempſtreſs with much more ſpeed and briſkneſs than thread of the ſame quality bleached in the field; it is leſs brittle, and, on that account, is better for the weft, as well as the warp. It likewiſe may be ſtruck much more effectually home to its place in weaving, and does not afterwards move. I received this valuable obſervation from impartial and unprejudiced manufacturers, for whom I bleached thread according to this method for making handkerchiefs.

Before I conclude the preſent chapter I muſt obſerve, that the conjunction of the old and new methods of bleaching may be of incalculable advantage.

advantage. For however great may be the ſpeed of bleaching by the oxygenated muriatic acid, it is ſcarcely poſſible to adopt it in an extenſive manufactory, to the excluſion of the method of expoſure in the field, without very heavy charges in workſhops, tools, and utenſils; I would therefore adviſe, that the entire bleaching, without expoſure in the field, ſhould be confined to ſuch goods as are intended to receive, what may be called, a half or three-quarters white; and that thoſe which require a higher bleaching ſhould be finiſhed by expoſure in the field. By this arrangement the production of each kind of white colour will be diſtributed, ſo as to be very ſpeedy, and to acquire the requiſite degree of perfection in a very economical manner for the manufacturer, under all the heads of time, expence, and labour. The high price which may be afforded by a piece of a perfect white, and fine quality, will be a compenſation for the price of common goods intended for ordinary uſe.

CHAP. XII.

Of the First Dreſſings.

IN order to give more clearneſs to the white colour of bleached goods, it is uſual to give them certain dreſſings. Fine piece-goods, ſewing thread, ſtockings, gloves, and other articles of thread, or cotton, are more particularly ſubjected to this treatment. The following inſtructions may be ſufficient to ſhew the management of theſe ſeveral articles, after they have been ſubmitted to the laſt immerſion. 1. The piece-goods are firſt to be preſſed, or wrung, in the ſame manner as after taking them out of the muriatic acid, and in this ſtate they are to be immerſed in water, rendered ſour by ſulphuric acid, to ſuch a degree that it may mark from two and a half to three e grees of the areometer of Moſſy. The Iriſh manufacturers, who uſe this acid in preference to ſour milk, for the bleaching their piece-goods, compoſe their bath of one hundred parts of water to one of acid. This proportion communicates to the water a taſte reſembling that of ſtrong lemonade.

The

The French bleachers, particularly thofe of Mayenne and its environs, who are accuftomed to pafs their piece-goods through fulphuric acid, compofe their bath of fixty pots of water to one pot of fulphuric acid, and they leave their goods immerfed therein during the whole night. The bath may be ufed cold, but it is more effectual and fpeedy in its operation when heated, and appears befides to throw out the colour to more advantage. If it be thought proper to ufe the heated bath, it will not at all be neceffary that it fhould exceed the heat which the hand can conveniently fupport. But it is adviseable in that cafe, to pour in the fulphuric acid at the time when the hot water is added, or elfe to mix it with one of the meafures of cold water, which may be ufed to dilute the mixture, or to cool it. At the time the hot water is poured out of the boiler, the acid muft be poured gently, and with care, becaufe it is liable to fly about *; and the greateft attention muft be paid to mix it well with the water, in order that the bath may be equally acid throughout.

The goods which are immerfed in the bath

* This danger is obviated by actually plunging the neck of the bottle in the water. The fulphuric acid immediately quits the bottle, in confequence of its fuperior weight.—*T.*

may

may be left for half an hour, or an hour, without the leaft apprehenfion of any ill confequence. I have even left them for twelve or fifteen hours without the flighest accident. The bath muft be left covered whether it be hot or cold, principally in order to prevent any accidental dirt, or other impurity, from falling into the fluid. The goods muft not be too much preffed together: the lefs they are preffed, the fooner the acid will produce its effect, and confequently the lefs time will be required for them to remain in it.

When the goods are taken out of the bath they muft be preffed or wrung, and then kept in a ftream, or large mafs of water; that is to fay, until upon rinfing them out in various parts, and applying the tongue to the part that is cleared, no acid tafte fhall be perceived. If any fuch tafte remain, the goods muft be immerfed again, if in a running ftream, or the water muft be changed if they be fteeped in troughs.

When the goods are ready for taking out, they are to be wrung with the winch, or hook, before defcribed, and preffed, and then paffed through the blue liquor if neceffary. The bluifh caft may be given in two ways, either by paffing them through a hot or cold folution of the

the ſoap, in which a ſmall quantity of fine indigo has been diffuſed in the uſual manner by means of a bag; or otherwiſe the piece may merely be plunged in a ſolution of white ſoap, without any kind of blue, while it ſtill contains a ſmall portion of acid. In the latter caſe, the acid contained in the cloth immediately developes in the bath a ſlight tinge of Pruſſian blue, ariſing from the particles of iron combined with the alkali of the ſoap, which tinge is diſtributed very equally on the whole piece. I muſt remark, that I have always uſed white veined, or mottled ſoap, to produce this laſt kind of blue. As a certain degree of experience is neceſſary in the uſe of this ſecond method, which, nevertheleſs, poſſeſſes the recommendation of economy, I would adviſe the operator to make uſe of the former, until, by trials, he is ſo far accuſtomed to the ſecond as to have no fear of miſtake.

The piece-goods being then well preſſed are always dried upon lines of hemp, or, which is ſtill better, of hair ſtretched on poles properly diſpoſed under cover; the pieces are held on the lines by the uſual wooden peg or clamp of the laundreſſes, or they may be hung over poles cleared of the bark, and covered with coarſe cloth, in order that the goods may not receive

any

any ſtain from the wood. It ſeems, upon the whole, a matter of indifference, whether the drying be performed in the ſhade, or in the ſun-ſhine.

With regard to common or coarſe cloths, which require no very extraordinary bleaching, they are never blued unleſs it be required. In every caſe, as it is eſſential that the blue ſhould be given with as much evenneſs as poſſible, care muſt be taken to paſs the pieces over a reel placed above the veſſel expreſsly appropriated to this purpoſe.

Linen or cotton thread, &c. is dreſſed in the ſame manner as has been directed for piece-goods. Theſe may be plunged in a bath of ſulphuric acid upon poles, or in the ſame manner as has been directed for the bath of blue. It is not neceſſary to turn them, becauſe the poles are fixed in the veſſel beneath the ſurface of the liquor; or, inſtead of this method, they may be diſpoſed in layers in baſkets of white willow. The bath of ſulphuric acid may be uſed for all kinds of goods, though in proceſs of time it aſſumes a ſlight amber colour; it is poſſible, nevertheleſs, to uſe it without danger until it is entirely exhauſted, taking care only to reſtore it from time to time, by the addition of that quantity of acid which may be neceſſary to keep up its

its ſtrength; when, however, the bath of acid has at laſt acquired too deep a colour, it muſt be entirely renewed.

Linen and cotton threads are likewiſe plunged in the bath of blue, after having immerſed them in the acid upon the poles; but in order that this part of the preparation may be diſtributed equally, the ſkeins muſt be turned once or twice half round; they may likewiſe be preſſed by the hand, provided the quantity be ſo ſmall as to render the economy of time an object of no conſequence.

Every kind of wood may be employed with nearly the ſame advantage, for the baths of ſulphuric acid or blue, without fear of ſpotting the goods. I have uſed oak, cheſtnut, white-wood, and deal. The latter, however, is preferable, if at hand. Great care ſhould be taken, that there ſhould be no nails in it; and before the pieces are plunged in either of the baths, the acidulated water muſt be well ſtirred, in order that the acid, or the blue, may be equally diſtributed. Stockings, night-caps, gloves, &c. of thread or cotton, require particular management. After the bath of ſulphuric acid, and before they are paſſed through the blue, it is always advantageous to give them a good ſolution of white ſoap, in which they muſt be well

 rubbed,

rubbed, for the purpofe of completely removing the fpots of oil or greafe of the manufacturer, which may have refifted the black or green foap, or the lees, to which they have been fubjected during the procefs of bleaching; for it very feldom happens, that they are entirely clear of fuch fpots, becaufe the dirt of the hands with which almoft all this kind of goods are covered, frequently prevent their being feen. The oxygenated muriatic acid having likewife little or no action on fat or oily bodies, the different objects fo fpotted might be plunged to little purpofe in that fluid.

Stockings, gloves, &c. when taken out of the folution of foap, are to be cleared in clean water, after which they are to be fubjected to the prefs, or wrung, previous to paffing them through a flight infufion of blue. The fecond method of communicating the blue tinge, as before defcribed, may here be ufed, that is to fay, they may be plunged in the folution of foap immediately upon taking them out of the acid, and rinfing them, if the goods have no fpot upon them, which it is effential previoufly to difcharge. After they have received the blue, they are to be preffed, or wrung, and dried upon cords, firft turning them infide-out, for fear of foiling the place of contact. This precaution

caution ought to be taken before they are paſſed through the blue, and at the time of taking them out of the ſolution of ſoap.

Linen or cotton thread requires its hanks to be opened and ſeparated, in order that it may dry more ſpeedily. This is the moſt certain method of preventing the entangling of the thread, and their conſequent breaking, which would not fail to happen ſooner or later, eſpecially with ſingle thread, if the operator were to ſuffer them to dry before they were ſeparated from each other. This ſeparation is to be made after they have been preſſed, or wrung out of the laſt fluid. It is time enough, nevertheleſs, to do it when they are half dry, and in this ſtate it is, in fact, rather more convenient.

The following method is very convenient to untangle the thread, particularly ſuch as is fine, and reſtore it to its original ſtate, when, in conſequence of the operations of bleaching, the ſkeins may have been mixed in ſuch a manner as to endanger their breaking, if an attempt were made to clear them by any other means. It conſiſts ſimply in extending each ſkein ſeparately, and ſlightly, under water. By turning and returning them, and afterwards extending them with the hand, the threads will very ſoon arrange themſelves, and obtain their original

 ſituation

ſituation gently, without obſtacle, and without giving any cauſe to apprehend their breaking.

If by accident the thread ſhould become dry, while too much intermixed to be cleared and wound off in this ſtate for wefts, warps, or other uſes, nothing is more effectual to clear the ſkeins, than ſlightly rubbing them with linſeed oil, here and there. This method is uſed for entangled ſilk; and I have found it ſucceed perfectly well with thread.

The operator muſt be aware not to ſoap cleared objects (ſuch as callicoes, or other linen, or cotton goods, dyed or printed) in the ſoapy ſolution, which has been uſed upon pieces taken out of a ſtrong oxygenated muriatic acid, even though they may have firſt paſſed through a bath of ſulphuric acid. For this ſolution of ſoap does not fail to acquire the property of bleaching and diſcolouring other goods, unleſs the goods which were ſoaped have, after being taken out from the muriatic acid, been immerſed in clear water for a long time. I have ſeveral times beheld this effect with ſurprize; I have frequently remarked, even that pieces, which after being taken out of the bleaching liquor, have undergone an immerſion in the ſulphuric acid bath, ſtill retained a ſufficient quantity of muriatic acid to aſſume a yellowiſh tinge

after

after they had paſſed through an infuſion of indigo. This laſt ſometimes aſſumes the ſame tinge after an interval of twenty-four or thirty hours.

To avoid this inconvenience, it is neceſſary not only to cleanſe the piece by rinſing after its laſt immerſion, but likewiſe to give it a ſlight immerſion in a ſoapy or alkaline water, from which it muſt be afterwards well cleared.

The bath of indigo muſt be compoſed in ſuch a manner, that it may not be neceſſary to reſtore it during the immerſion of the ſame article; otherwiſe there would be a danger of its receiving different ſhades of blue.

There is much leſs danger of this in uſing the azure blue (powder blue of the market), the different ſhades of which are all previouſly prepared, and ſold in this ſeparate diſtinct ſtate. In either caſe it is neceſſary to plunge the goods in the bath, at the moment when the colour is ſuſpended, and to leave them in it no longer time than is neceſſary for them to imbibe it.

I have before recommended, that the waters of the immerſions, or bleaching liquor, ſhould be preſerved, even though too weak to act ſenſibly upon the pieces already in part bleached. They are uſeful in this part of the proceſs; that is, after the dreſſing with ſoap, for ſuch articles as are not to be passed through the blue: for this last dressing is not agreeable to every one, because it gives a greyish tinge to such parts as are not of a very firm white. After the immersion in soap-water, and the subsequent rinsing, the goods, being first well pressed, are thrown into this exhausted bleaching liquor, where they soon acquire a clearness, which has a very good effect. After they have remained in this fluid for half an hour, they are to be pressed and dried as before directed. It may also be remarked, that the same reserved bleaching liquor, though exhausted, is excellent for clearing and rinsing thread and stockings from their lixivium when they are already bleached. If it were applied to no other use, it is preferable for this use to common water. The several articles are much more speedily cleansed, and acquire a certain degree of improvement in the general appearance of their colour.

Some persons require in their goods a certain dressing, as it may properly be called, which affords, particularly to such as are of open texture, an appearance of firmness, which they lose when folded. The dressing may be given in a more or less durable manner. The first method consists in drying the goods, with scarcely any wringing, and hastening the drying

as

as much as poſſible. This dreſſing is, as may eaſily be imagined, one of the moſt innocent; but its effect diſappears by handling, or carriage, or by one ſingle time of wear. The ſecond dreſſing, which is permanent till after waſhing, conſiſts, as all the world knows, in incorporating ſtarch with the powder-blue for ſuch objects as require it, or uſing it without blue for thoſe which do not. The doſe muſt be varied according to the quality and kind of the goods, and the choice of the proprietors.

There are likewiſe certain articles, to which a greater or leſs appearance of firmneſs is given, by a dreſſing of glue or gum-arabic, which is mixed with the ſtarch after both have been boiled ſeparately.

There are alſo certain articles, ſuch as linens, which are frequently dreſſed with a decoction of rice.

Theſe different goods are always hot-preſſed by means of a cylinder; which operation gives them the proper degree of firmneſs.

As every object which can be of uſe to accelerate the work, and diminiſh labour, is of great importance, I ſhall here deſcribe the machine with which the Engliſh, who are the inventors, waſh their fine linens.

It conſiſts in a kind of churning inſtrument

(ſee Plate II. fig. 15 and 16), the circular part of which has four holes bored in it, into which are fixed pins of white-wood, rounded at their extremities. They are more or leſs long, according to the depths of the troughs in which they are uſed. The handle of this inſtrument is a croſs, or T, with which the pins at the other end may be moved in different directions, and uſed to agitate the ſtockings, and other ſmall articles depoſited in the troughs, containing the ſolution of ſoap. This inſtrument being moved in various directions, is very convenient in cauſing the ſuds to lather, and to impregnate the linen with great expedition. It appears to me, that ſuch an inſtrument may be uſeful to cleanſe ſtockings, gloves, and other articles of cotton or linen thread, which may be required to be bleached or diſcoloured, as I ſhall hereafter more fully explain.

I have been informed by an Engliſh woman, whom I ſaw make uſe of this machine, that they have others of the ſame kind in England, of a ſize adapted to common waſhing; and even ſome of ſuch large dimenſions as to be moved by horſes. If this be practicable, it might, at leaſt, be worth trying. And for this reaſon it is, that I have thought proper merely to deſcribe the ſmall inſtrument I have myſelf ſeen.

The

The ſame recommendation may be offered in favour of another machine, which is uſed in England to rub coarſe linen. It conſiſts ſimply in two ſtrong planks with grooves, the uppermoſt of which is moveable; the motion which it gives to the cloth paſſed between them cauſes the ſoap, or lees, in which it is plunged, to lather, and contributes to clear it of its impurities.

The Engliſh likewiſe uſe, with advantage, for waſhing clothes, two grooved cylinders, running one upon the other, by means of a handle above the veſſel of water, between which cylinders a number of pieces of cloth are paſſed at a time, their extremities being ſewed together, ſo as to form a long loop, or endleſs web. A few turns of the cloth between the cylinders (the lower of which is abſolutely covered with water) are ſufficient to cleanſe it from all the impurities which the lees have opened and detached. I have not thought it neceſſary to give a deſign of theſe two machines, which are executed and uſed with ſucceſs at St. Denis, as well as at Beauvais.

CHAP.

CHAP. XIII.

Of the last Dressings.

PIECE-goods, bleached by the oxygenated muriatic acid, if left to themselves during the course of bleaching, are subject, from the nature of their thread and texture, to swell and contract; and, consequently, to lose, in their dimensions, particularly in length. It is essential, therefore, not only in order to recover this measure, but likewise to render the piece more uniform, softer, and more equal in its grain, that it should receive a proper dressing, to restore its original state. The necessary treatment for producing this effect, is by means of an apparatus described in Plate VI. fig. 1 and 2. It consists of a light frame of wood-work, on the upper part of which there are several light bars fixed across, very near one another; and on each side is a grooved piece. The piece-goods are drawn between these bars and the stretcher; after which they are rolled on a wooden cylinder, moved by wheel-work, turned by one or two men, according to the degree of tension required

quired to be communicated. This dreſſing on the roller may be performed either with or without the application of heat; the piece being either humid or dry.

It muſt be dreſſed in a dry ſtate, without heat, when nothing more is required than to ſoften the grain, and to reſtore its original dimenſions. On the contrary, the operation is performed with the aſſiſtance of heat on the humid piece, where it is required not only to be ſtretched but dried at the ſame time. Independent of the force of tenſion to which the piece is ſubjected in the direction of its length, the perſon employed to roll it upon the cylinder muſt be careful to pull it out by the ſelvedges, to the ſame width as that part which is fixed upon the cylinder already. The piece muſt then remain at leaſt twenty-four hours upon the cylinder, that the folds, or wrinkles, may be totally effaced, and its original dimenſions permanently reſtored. The rollers, or cylinders, ſhould be covered with cloth, to prevent the bleached goods from being ſoiled, and iron plates muſt be placed beneath for the purpoſe of drying them, when it is required that this operation ſhould be performed at the ſame time as they are ſtretched and rolled.

When the object of the manufacturer is ſimply

ſimply to take out the folds or wrinkles of the piece, without altering its grain, it is to be ſucceſſively paſſed over ſeven wooden rollers, diſpoſed one over the other in ſuch a manner, that by turning one, the ſix others, over which the piece is rolled, muſt likewiſe turn. The handle of this apparatus is fixed to the lower roller, and requires one perſon to turn it. I have not thought it neceſſary to make a drawing of this machine, as it is now uſed in many manufactories and workſhops.

When it is required to compreſs the grain of the piece, or to glaze it, it muſt then be paſſed through a hot calendar, ſee Plate VII. fig. 1, 2, 3, which conſiſts in a ſolid frame of wood-work, in which moves a braſs cylinder, kept at a certain degree of heat, by means of balls or bars of iron heated in a furnace, prepared for the purpoſe. This cylinder is placed between two others of walnut-tree, and of twice the diameter. Before its arrival at theſe cylinders the piece is paſſed over and under ſeveral bars, as well as through a ſtretching frame. This machine is uſually moved by a horſe, though there are ſome which are worked by hand, but, of courſe, with leſs expedition. In either caſe, that is to ſay, whether they may be intended to act with or without heat, although the former is preferable,

able, the machine is always ſet in motion by a train of wheel-work, to procure a ſomewhat greater degree of tenſion, and more perfectly to efface the folds. When the piece paſſes under the heated cylinder, it is to be ſlightly ſprinkled with water, by means of a ſmall broom or rod. In ſome bleaching works they uſe the mangle, more or leſs loaded; a well-known engine, conſiſting of a ſquare heavy box, which is made to run backward and forward upon cylinders of wood. This machine, which is commonly moved by water, or by a horſe, does not glaze the goods, but lengthens the meaſure. It neceſſarily implies the uſe of an engine, like that deſcribed in Plate VI. to diſpoſe the pieces upon the cylinders. The mangle is alſo neceſſary when the goods are required to be damaſked, that is to ſay, when the effect of a wave is deſired upon their ſurface. This laſt dreſſing is eaſily obtained, by rolling the folds a ſlight degree in zig-zag.

The following method is likewiſe very much uſed for drying piece-goods by the dealers of thoſe articles. It is extremely ſimple, and conſiſts ſimply in a plate of copper or braſs placed on an iron ſtand, under which is placed a baſon of charcoal, or burning embers. The piece being paſſed over this heated plate, dries gradually,

ally, and very ſpeedily. This method is equivalent to a ſecond paſſing through the ſtretching engine, and ſerves well for drying, but does not reſtore the original meaſure. It is, however, perfectly well ſuited to ſtockings night-caps, &c. The heated plate ſhould be kept perfectly clean, leſt it ſhould ſoil the goods.

The method of drying, and giving uniformity to piece-goods, being deſcribed, it now remains to be ſhewn, how thread is to be arranged and dreſſed after drying on the pole. The preparation given to this article tends to clear the thread from that roughneſs which it acquires from handling, and never fails to exhibit when dry. This is eaſily removed by ſhaking each hank, either upon the pin or the hand, after having rubbed it between the hands, or beaten it with a mallet. The operation is particularly neceſſary for ſkeins of ſewing ſilk, which, as I have before remarked, are diſpoſed to curl up and ſhrink from their original meaſure; to which they may, however, be very ſpeedily and conveniently reſtored by means of a kind of frame, ſee Plate VIII. fig. 3 and 4, acroſs which the ſkains are to be paſſed and ſtretched. One of the moveable croſs-pieces is to be raiſed and fixed, by means of the pins which enter into holes diſpoſed in a zig-zag direction, at a ſmall diſtance

diſtance from each other, on the apparatus of the frame. It may eaſily be underſtood, that each ſkein being moiſtened and wrung, and afterwards paſſed over theſe croſs-pieces, muſt remain thus ſtretched for a certain time, that is to ſay, until it is dry, and by that means forced to preſerve the length, which it has received from this tenſion. But this laſt preparation is not uſually given, except to double or ſewing thread, which muſt be ſpread out as much as poſſible upon the croſs-piece, in order to haſten the drying.

The ſkeins of ſingle thread are afterwards to be tied up in the middle, and put together by ſcores, or quarters of the hundred, in a bundle, tied together with a ſkein of the ſame thread. At leaſt the finiſhed thread, in ſome provinces of France, is thus made up for delivery from the bleacher to the merchant or manufacturer. With regard to ſewing or double threads, they are turned up in a ſpiral or twiſt, and, in order that they may lie cloſe, this operation is made on the pin. See Plate II. fig. 12. This, in France, is called folding up in carrots.

Stockings, night-caps, &c. of thread or cotton, as ſoon as dry muſt be examined, to take up the ſtitches, or repair them, for it very ſeldom happens, particularly in ſlight goods, ſuch as ſtockings,

ſtockings, that they paſs through the various operations here deſcribed without ſome ſtitch falling; there is, indeed, little to be feared if the cotton be knitted by hand, but moſt goods which are wove in the ſtocking engine, which is employed in preference for articles intended for ſale, are very unequally knit; and many articles are made, like the Engliſh goods of the ſame kind, with two or three threads, which diminiſhes their ſtrength ſtill more. Stockings, gloves, &c. after examination, turning, and repairing, if neceſſary, are diſpoſed in the preſs, folded in three folds according to their length, then ſorted according to their quality, faſhion, and dimenſion, in ſixes or half dozens. After dreſſing, if required, they are then put into the preſs; and laſtly, wrapped in blue or white paper, and properly marked.

Theſe laſt dreſſings are not commonly given, unleſs the employer requires them; otherwiſe theſe kind of goods are returned as ſoon as dry, even without turning them, for fear they ſhould be ſoiled in the carriage. In many places ſtockings are dried on the leg, in order that they may recover the contraction they have undergone in the bleaching, for it is the property of frame-work knitting to ſhrink and draw up a little when wetted. No more than one pair of

ſtockings

ſtockings is put on each leg; and to prevent their ſhrinking, as they dry, the upper part of the leg is fixed with pins after the ſtretching has been carried to the required extent. The forms, or legs, ought to be made of aſh-wood, and, if poſſible, of a ſingle piece, becauſe the ſtockings may be torn at the place where the two pieces that form the foot and leg are joined together, as is frequently the caſe with theſe implements. The corners ought to be very carefully taken off, to prevent the ſame accident from taking place.

Stockings, whether of thread or of cotton, are likewiſe ſinged with a hot iron, which is paſſed over the article, properly ſtretched on the leg. This preparation is not to be given but juſt before the laſt bathe in the lixivium, or immerſion in the acid, on account of the reddiſh brown colour, which is the conſequence of this proceſs, and requires to be cleared off.

Inſtead of this proceſs with the hot iron, the goods are ſometimes paſſed over a lamp of burning ſpirit of wine. Theſe particular dreſſings are only given to ſuch goods as are required to have a very uniform appearance, and the moſt exquiſite white, in imitation of goods of the ſame kind which we import from England.

Having thus ſhewn the method of dreſſing, ſquaring, and drying the pieces, I ſhall now proceed to give directions for folding them. This operation is uſually performed on the ſtick, becauſe it is very expeditious, and the goods are in this way very properly folded.

Moſt piece-goods are folded in two, acroſs their length (ſee Plate II. fig. 7); for this purpoſe, one of the ends of the piece is paſſed over a moveable roller, previouſly ſuſpended at each extremity in the loop of a cord, fixed to the cieling of the work-ſhop: this length of the cloth is thrown on the other ſide of the roller, and the workman continues to make the fold he has began, by drawing what he has folded equally over the roller. In this manner the piece becomes completely folded up. Attention muſt be paid, that it ſhould fall upon a ſtage, or board, in order that it may not be expoſed to injury, or dirt.

The cloth, thus folded in two, is carried to a table (ſee the ſame Plate, fig. 8 and 9) to be folded in this ſtate, in equal and regular folds. The length of the intended fold is taken with the piece itſelf, which is applied to two flat iron bars, fixed on each ſide of the table, and pierced with different holes, to fix the ſupports which determine the length of the folds; after which,

which, with a firſt rod reſting on the two firſt ſupports, the cloth is thrown over the rod, and thence carried to the ſecond, which is to form the oppoſite fold. In this place a rod is laid to form the fold, and the cloth is again carried to the oppoſite extremity, where a ſecond fold is made over another rod. Hence the workman proceeds, as before, to the other extremity, and the ſame proceſs is thus repeated to the end of the piece, drawing out thoſe rods, as he proceeds, which were firſt laid. It is not neceſſary to uſe more than four or five rods at each ſide. They are of poliſhed iron, of the thickneſs of a little finger; the pins, or ſupports, placed in the holes of the flat bars of iron, ought to be ſufficiently long to contain all the folds which the cloth may require. The length of theſe folds are proportioned to the extent, or volume, which the piece is expected to occupy after the folding. Experience will ſoon direct the operator in this matter.

To this operation ſucceeds that of the preſs, if the pieces are thought capable of it, with regard to their bulk, and the facility in diſpoſing them for that purpoſe. On taking the piece out of the preſs, when it is folded in equal folds, whether ſingle or double, it is turned inwards, ſo as to form one general fold, as may be ſeen in

Plate II. fig 10; in which ſituation it is ſecured by ſtrings of different ſize, according to the quality of the piece. The piece thus packed together is ornamented with taſſels of gold thread for fine goods, ſuch as cloths, cambrics, muſlins, or of ſilk of different colours for more common goods. With regard to coarſe goods, it is uſual to knot the ſtrings together in the front part of the fold.

I muſt here obſerve, that piece-goods loſe in their length by bleaching about one ell, or an ell and a quarter in twenty-five, according to their quality; and this loſs is reſtored to them again by the laſt dreſſings here deſcribed.

With regard to the loſs of weight which threads undergo, it depends much on the rotting of the flax, according as it has been more or leſs perfect. The linen threads of Flanders and Artois, for example, which are rotted in water, do not loſe more than 20 per cent; whereas thoſe of Picardy, of which the rotting in water is ſo far from ſufficiently waſhing the thread, that it acquires, on the contrary, a degree of impurity from the earth on which it repoſes, in addition to that which is detached in the courſe of time by the maceration, which its external part has undergone, loſe more than one fourth of their weight, generally ſpeaking. Coarſe threads naturally

turally lose more than others. In general the loss may be estimated at twenty-five or thirty per cent, and eighteen or twenty-five for those of middling quality. With regard to cotton threads, the loss is scarcely three or four per cent. Cotton piece-goods may lose more in proportion, on account of the dressing which was added to their weight, and which must first be dissolved, and taken out of the cloth, before it can be prepared for the discolouring or bleaching process.

Having treated of the ordinary dressings used in France, I think that the reader will receive, with pleasure, some account of the singeing, particularly used by the English for light cotton goods. Every one knows that muslinets are striped, plain, and spotted; muslins are more beautiful in proportion as they are less downy, or covered with fibres of the cotton wool. On this account the English, who are likewise attentive to use cotton of long staple for these goods, take the greatest care to render them as smooth as possible; this may be observed, particularly in their frame-work, and other cotton goods, of which the beauty of the bleaching is the more conspicuous, in proportion as the surface of the article is smoother, or less covered with the light down, which is observable on all the articles of cotton when firſt manufactured.

I have, therefore, preſumed, that my countrymen will learn with pleaſure how the Engliſh manufacturers contrive to take off all this down, which on a beautiful and fine piece of cloth is ſingularly hurtful to the reflection and brightneſs of the white colour it has received. With this view I have given a drawing of the machine uſed for this dreſſing. See Plate VIII. figs. 5 to 10, and the deſcription. It will be ſufficient to obſerve in this place, that after having ſewed the muſlins to the coarſe cloths, which are nailed upon two rollers, with handles to ſtretch them, they are rubbed with a bruſh to raiſe the down. The bruſh is paſſed once or twice over the face of the piece, after which a bended bar of iron, more or leſs ignited, is ſpeedily and lightly paſſed over the upper ſurface. This bar, according to its degree of heat, is paſſed two or three times over the ſame place, and after it has been repeatedly moved along one border, it is inſenſibly moved towards the other. When the down of this firſt length is well cleared off, which is obſerved by looking aſlant upon the ſurface of the piece, a new portion is ſtretched, by rolling another part of the piece, which is to be treated as before.

Piece-

Piece-goods which are intended to produce an effect on both ſides, are ſinged on the back as well as the face, but more ſlightly on the former than the latter. It is neceſſary to have at leaſt two or three irons, one of which is to be heated, while the other is in uſe; and the greateſt precaution muſt be had to make them very clean previous to each time of uſing; this is done upon a rag, or a fine grained ſand-ſtone, when they are taken out of the furnace in which they were heated. This attention is neceſſary, for fear ſome greaſy ſubſtance, or tallow, might attach itſelf to the face of the iron, and burn, or penetrate the piece. The irons may be heated indifferently with turf or pit-coal, as well as with wood.

Cotton cloths, or muſlin, which are rendered even and ſmooth by this method, immediately acquire by this treatment, which is always performed at the commencement of the bleaching, a brown tinge ſimilar to that exhibited by linen burned in ironing, but this colour moſtly diſappears at the firſt or ſecond immerſion, without any intermediate lixiviation.

This management with regard to cotton goods, or muſlins, is equally applicable to linens, though theſe are leſs ſubject to the downy

 covering,

covering, on account of the length of the ſtaple of which they are compoſed.

It is very poſſible to uſe the ſame proceſs with ſtockings, night-caps, and other fine knit works in cotton, capable of being rendered more valuable by their clear white colour, which always ſeduce the conſumer, who is, for the moſt part, leſs attentive to the quality of the merchandize than its external appearance But I have already remarked, that the mechaniſm to diſpoſe knit, or frame-worked goods, to receive the ſingeing, ought to be different from that of piece-goods, and appropriated to the form of the object.

There is another method alſo of ſingeing cotton goods and muſlins, namely, by ſpirit of wine; but this method never operates with the ſame ſmoothneſs and equality as the red-hot iron, and is alſo much leſs expeditious. Nevertheleſs, as it may be uſeful and applicable to certain goods, the reader may conſult Plate VI. fig. 1 and 2, where I have deſcribed the machine which may be uſed for this purpoſe. I need only remark, that inſtead of the box which contains the hot embers, another muſt be placed, containing a row of wicks for burning ſpirit of wine. One man is ſufficient to attend

tend and direct this method, as well as the former; and the ſpirit of wine made uſe of may be mixed with a certain proportion of brandy, or otherwiſe it may be lowered in its ſtrength, as the operator may find beſt ſuited to his purpoſe.

CHAP.

CHAP. XIV.

Concerning Proof Liquors.

BY proof liquors, I underſtand all thoſe coloured fluids, which are extracted from vegetables by decoction or fermentation, and which, when mixed with the oxygenated muriatic acid, are more or leſs changed, according as the one or the other liquors is more or leſs concentrated: theſe vegetable fluids, according to the degree of alteration which they undergo, are of uſe to aſcertain the ſtrength, and more particularly to afford a judgment of the uſe to which the muriatic acid may be applied, when after having been prepared according to the directions already pointed out, it has been applied to one or more immerſions. It is true, indeed, that this acid might be more concentrated by putting leſs water into the pneumatic veſſel, or by increaſing the doſes of the ingredients; but this would afford no advantage excepting for the purpoſe of carrying it from place to place, or for the more ſpeedy bleaching of coarſe goods, or other objects of the ſame quality; ſuch as towel pieces, coarſe thread,

thread, twifts, and the like, of which there is no reafon to fear an alteration in their texture. For every other merchandize of a fine quality, it is always prudent to dilute the acid in a certain proportion for its moft advantageous application.

A folution of one part of indigo in eight parts of fulphuric acid, is particularly pointed out by Berthollet as having been ufed by De Croifille at Rouen. This preparation differs from a folution of Saxon blue in no other refpect than that this laft is made with one part of indigo to four parts of the fulphuric acid. Either of thefe compofitions may be digefted in a matras, or fimply in an apothecary's phial, placed on a water or fand bath, or in hot afhes: at the expiration of feveral hours part of the indigo, previoufly pounded and fifted through filk, becomes diffolved. This folution, which is of a very deep and denfe blue, is to be gently poured by inclination into an appropriate veffel, after which it is to be diluted with water, until it marks one degree below zero on the areometer of Moffy. In this ftate it forms a proof liquor, of which three parts will be rendered colourlefs by one feventh or eighth part of oxygenated muriatic acid, compofed in the manner before defcribed. This liquor may be meafured

in

in the cap of the case of the areometer, and then poured into a glass cylinder, which is graduated and stands upon a foot. See Plate IX. fig. 10 and 11.

I have thought proper in this place to mention the kind of measure which is made use of in this proof, because the degree of strength indicated by the fluid is very frequently different when the oxygenated muriatic acid, is poured into a vessel of a different diameter. It is therefore essential to use the same measure constantly.

It must be observed, that the oxygenated acid may be used to advantage, as a second bath for linen goods, already in the progress of bleaching, or as a first bath for cotton goods, from the time its strength is such, that one single half part of the acid is required to render three parts of the blue fluid colourless, until that state in which it requires one entire part of the latter to discolour three of the blue. When the bath is weakened to this degree, it is no longer applicable but to the preparation.

Nevertheless, if there be a certain quantity on hand, it may be used for steeping and preparation, in case there is time for such goods as are either cleared or uncleared; for though the acid be so weak that it does not seem capable of bleaching, nevertheless it will act in process

o

of time, as long as it is capable of difcolouring the proof liquor. For I have many times remarked, that however weak the preparation may be in which the piece is thus immerfed, the white colour of this laft does not fail to be very perceptibly forwarded, when it comes to be again fubjected to the action of a newly-prepared and ftrong bleaching liquor. This exhaufted fluid muft not, therefore, be rejected, even though one whole meafure fhould be required to difcolour one meafure of the blue folution of indigo, with which it may be mixed. In this manner trials may be made to afcertain its ftrength, as foon as it is weakened to that degree that three parts of the blue liquor are difcoloured by one of the acid. The operator cannot pay too much attention to the total exhaufting of the power of the bleaching liquor, fince, upon the whole, thofe weak folutions may be very profitably applied in a well-regulated manufactory. When the liquor of the bath no longer acts upon the proof liquor, it is entirely exhaufted of the oxygenated muriatic acid, though it ftill preferves a peculiar odour, which is not difagreeable; it then confifts of the common muriatic acid diluted with water, if it be the refidue of a bath of the odorant muriatic acid; but, if, on the contrary, the fluid be the

refidue

refidue of the oxygenated muriatic acid without fmell, it contains likewife a combination of that acid with pot-afh. In either cafe the fluid may be thrown away, if the operator is not aware of any peculiar purpofe to which it may be applied; if he has fuch a purpofe in view, he may referve it either for that object, or for rinfing fuch goods as are already bleached, and have paffed through the laft lees, as has before been remarked, for which laft purpofe it appears preferable to common water.

The tincture of cochineal may be ufed as a proof liquor, according to the information given in the annals of chemiftry. It is equally valuable with the Saxon blue, and even made with lefs trouble; nothing more being required than to boil a fmall quantity of the cochineal, firft crufhed in a marble or glafs mortar, or ftrongly rubbed between the fingers; the decoction muft then be filtered through cloth, or blotting-paper, upon which it muft be poured flightly by inclining the veffel, in order that the remains of the infect may be feparated from the fluid.

Two parts of the tincture of cochineal diluted with water to half a degree below zero, require two parts of the muriatic acid without fmell, at the fame degree as that which is neceffary to render the blue proof liquor colourlefs. The

tincture

tincture of cochineal becomes converted to a yellow colour.

It may here be remarked, that the violet liquor of Javelle concentrated to four degrees of pot-ash below zero, requires no more parts to discolour three parts of the blue before described, than are required of oxygenated muriatic acid to discolour the same quantity of blue; but the liquor of Javelle bleaches more speedily and uniformly.

The tincture of turnsole may also be used with no less advantage than cochineal, and is prepared in the same manner; and, lastly, the tincture of beet-root, and even wine itself, may be used with equal convenience, if other vegetable tinctures and decoctions be not at hand. The juice of acacia and currants are also susceptible of affording the same indications.

With regard to the colours which the different tinctures assume, they are as follows:

The Saxon blue, or solution of indigo in the sulphuric acid, becomes a yellow, more or less inclining to fawn colour, whether it be mixed with the oxygenated muriatic acid with smell or without. Its tint becomes deeper in proportion as the blue is more intense.

The tincture of cochineal assumes an orange colour.

Red

Red wine of Macon acquires an opal colour; the infuſion of turnſole becomes of a light amber colour with either of the acids, prepared in the manner recommended in this treatiſe; I have, nevertheleſs, obſerved, that it does not change with the fluid called the lixivium of Javelle *(leſſive de Javelle)*, which nevertheleſs cauſes a ſlight effervescence with vinegar.

It is very eaſy for the operator to regulate his proceſs with regard to every kind of tincture or infuſion, which he may find it moſt convenient to uſe, whether of woods or roots, according to the preparations I have laid down for cochineal and indigo. The latter ſolution may be prepared without the aſſiſtance of heat, as follows. After having poured the proper quantity of ſulphuric acid into an earthen or ſtone ware bottle, the pounded and ſifted indigo is poured in, and ſtrongly agitated by the hand for an hour, which is a ſufficient time for the clots of indigo, which are formed either at the ſurface of the acid, or on the ſides of the bottle, entirely to diſappear: during this agitation a ſtrong effervescence takes place in the fluid. The indigo, when well ſhaken, and penetrated by the acid, is ſoon diſſolved, but it uſually requires half an hour for that purpoſe. This portion of time is applicable to two ounces of indigo in

one

pound of the ſulphuric acid. To this quantity of acid half a glaſs of water may be added. I am convinced by experience, that this addition renders the action of the acid upon the blue more ſpeedy.

CHAP. XV.

The Methods of remedying ſuch Accidents as happen during the Courſe of Bleaching.

THE accidents likely to happen in the courſe of bleaching, may be diſtinguiſhed into accidents *of the diſtillation*, accidents *of the immerſion* in the alkaline or acid liquor, and accidents *of the dreſſing*. I ſhall give a ſhort account of theſe three clauſes of accidents, and at the ſame time point out their remedies.

Accidents in the diſtillation. The principal accident which is capable of interrupting the diſtillation, is when the lutes of the adopter ſuffer the gas to eſcape. The moſt ſpeedy remedy, in this caſe, to prevent the exhalation of the acid, which cannot be retained but with great difficulty, ſeldom for any length of time, and then very imperfectly, in conſequence of its great expanſion; the ſhorteſt method, I ſay, at leaſt if the diſtillation be not near its concluſion, is to remove the fire immediately from beneath the capſule of the retort, and to ſuffer this laſt to cool for a certain time, by raiſing it a little in its

its ſand-bath. If it be not poſſible to take it out of the furnace, together with its capſule, on account of the heat, or its ſticking too faſt, the adopter muſt be unluted from the funnel of the leaden tube, and the aperture of this tube cloſed with a cork, or lute, to prevent the gas of the pneumatic veſſel from evaporating; after which the retort muſt be raiſed, and placed gently upon a bag of ſtraw, or on coarſe cloths folded together; and then holding the retort by its neck, near the flexure, the adopter muſt be entirely unluted, by twiſting it round and drawing it off. The orifice of the neck of the retort is then to be cloſed with a cork ſtopper, but not ſo cloſely but that a very ſmall portion of gas may be ſuffered to eſcape, for fear of an exploſion. The ſtopper of the neck may, for greater ſafety, be ſlightly raiſed. This precaution is neceſſary, on account of the great expanſion of the muriatic acid gas. The old lute muſt then be taken off, as well from the adopter as the retort, and the places to which they were applied muſt be well cleaned, in order to receive freſh lute, after having carefully wiped off the moiſture with a cloth or a ſponge. If the lute which comes off be ſtill good, it may be kneaded again, adding, if required, a ſmall quantity of boiled oil, or it may be mixed with new lute, if it be burned or decompoſed. This decompoſition in the fat lute may be known by the white or reddiſh colour which it acquires, and the facility with which it breaks, on account of its having loſt the gluten which afforded it that toughneſs and tenacity, on which its goodneſs chiefly depends.

With regard to the lute of linſeed cake, it muſt, in almoſt every caſe, be totally renewed, particularly when internally applied, becauſe the heat hardens it too much to admit of its being kneaded again, with any moderate degree of facility; the decompoſition of this lute is known by the yellow colour it acquires, and the contraction it undergoes from the effect of the heat. The lutes being kneaded to a proper conſiſtence, and duly placed according to the directions laid down in Chap. IV. the adopter is to be fixed, previouſly removing the ſtopper from the mouth of the retort, and placing another in that of the ſmall end of the adopter, to prevent any inconvenience from the vapour which might iſſue out during the time of fixing it. This vapour is likewiſe condenſed within the adopter, in conſequence of its coldneſs. The retort is then to be placed, as before, on the furnace,

furnace, the adopter uncorked, and its beak luted into the tube of lead; after which, the fire is to be replaced beneath the capſule, and the diſtillation very ſpeedily recommences, and proceeds as uſual. This operation is a work of ſome delicacy; it requires to be performed with ſpeed, and great care muſt be taken while placing the lutes and the adopter in their proper ſituations, to ſtand always in ſuch a poſition, that the current of the external air may drive the vapour from the operator himſelf.

If the accident here deſcribed ſhould take place towards the end of the diſtillation, as it may ſometimes happen, in conſequence of the ſtrong heat which, at that time, may ſoften the lutes, it will be ſufficient if the fire be taken from beneath the capſule. The diſtillation ſoon ceaſes when this is done, particularly if care be taken to condenſe the gas, by the prudent application of wet cloths on the neck of the retort, as well as the adopter.

This inconvenience would not take place, if the workmen in thoſe glaſs-houſes which are principally employed in the fabrication of chymical veſſels could make retorts with necks recurved in the form of the adopter. Theſe kind of veſſels may be aſſiduouſly ſupplied by making ufe of a tube of lead, fo formed as to ferve inftead of the adopter, as I have already obferved, with regard to the tubulated bottles or bodies*. If, by accident, the lute which is adapted fhould fail, or fuffer the gas to pafs through, it may eafily be ftopped, by applying new lute to the place of junction. Inftead of the leaden tube, we may fubftitute, with ftill greater convenience (the danger of breaking excepted), a tube of glafs, of which the end neareft the bottle, or tubulated body, fhould be ground with emery. By thefe means there would be no application of lute, and confequently no danger to be feared with regard to the filtration of the gas, the efcape of which is eafily perceived by the fmell which diffufes itfelf through the workfhop, and is more particularly perceived when the nofe is applied near the veffels, or the lute. But as this laft method of difcovering the place where the lute has failed may be attended with the moft ferious confequences, if the greateft precaution be not ufed, it is more prudent to apply an open bottle of ammoniac near the

* This laft method appears to me preferable to every other, becaufe it requires only a flight attention to the lute, and can never produce thofe dangers which arife from the ufe of retorts.—*C.*

fufpected

ſuſpected place; at the inſtant that it is preſented, a white fume is formed, which immediately points out the defective ſpot. The bottle muſt be preſented above the current of air, which takes place near the lute, or in the workſhop. If this precaution be not attended to, the operator might be induced to remove a good lute, inſtead of one which was really defective.

On the other hand, if in the courſe of the diſtillation, and for want of keeping up the heat, the fluid in the pneumatic veſſel ſhould be abſorbed and riſe into the diſtilling apparatus, it is neceſſary the inſtant it is perceived to withdraw for a moment the ſtopper out of the neck of the retort, where, as I have already had occaſion to obſerve, the abſorption inſtantly ceaſes. Nevertheleſs, if, for want of being obſerved in time, the water ſhould riſe ſo far as partly to fill the retort, or body (for it never entirely fills it), the diſtillation will be ſtopped, from the coldneſs of the water, and its too great quantity. The ſhorteſt remedy is to draw out the exceſs of water, which is thus introduced into the diſtilling veſſel, by the aſſiſtance of a glaſs pump, or ſyphon, and afterwards to heat the ſame veſſel, firſt returning the water into the pneumatic veſſel, if thought expedient: but if the diſtillation be properly attended to, this accident can never happen.

Accidents in the lixiviations and immerſions. I place the accidents ariſing from theſe two operations in the ſame claſs, becauſe they can ſcarcely take place, but by the joint operation of both.

Any article which is badly cleared of the lixivium, and afterwards immerſed in the oxygenated muriatic acid, becomes almoſt immediately of a nankeen colour, particularly in the folds, either in ſpots where certain parts have not been ſufficiently rinſed, or elſe the colour is general, if the whole has not been well rinſed.

The ſame accident happens if ſoot has fallen on the linen or thread. The difference is ſimply in the colour, which approaches more to brown. Theſe colours are capable of becoming more and more deep if the miſmanagement be not remedied as ſoon as perceived, and that before the goods are ſubjected to other immerſions in the alkaline lees, or of the oxygenated muriatic acid. The ſame acident is to be expected, if the goods, though white at the time of their immerſion, are ſuffered to remain too long in the bleaching liquor. This does not fail to happen, particularly if the articles which are ſuffered to remain even in a weak ſolution, are kept in that ſtate the whole night. The next day they are found to be yellow, or charged with lixivium.

The

The remark which has here been made, concerning the nankeen colour, takes place alſo with regard to thoſe articles which, though white, have been immerſed in an exhauſted lixivium, or lees which have been uſed ſeveral ſucceſſive times for the immerſion of goods taken out of a ſtrong muriatic acid, without previous rinſing. Theſe articles, when taken out of ſuch exhauſted lees, and plunged into a new acid ſolution, undergo a change even though they may have been properly rinſed. I muſt in this place remark, that a lixivium may be exhauſted in conſequence of its combination with the muriatic acid from the goods which are plunged in it, though they may have been preſſed or wrung at the time of taking them out of the acid. Such exhauſted lees may, nevertheleſs, indicate a certain degree of ſtrength by the areometer, though in fact they do not poſſeſs it.

The only method of remedying theſe accidents conſiſts in the uſe of water, ſlightly acidulated with ſulphuric acid, no matter whether cold or hot, but the hot ſolution operates more ſpeedily. The ſpotted or tinged goods are to be ſoaked in this water for a few minutes, or a quarter of an hour, accordingly as the colour may be more or leſs deep, in conſequence of a ſeries of lixiviations

lixiviations or immerſions, more or leſs repeated. In this ſituation the offenſive colour is ſeen almoſt immediately to diſappear.

Inſtead of making a ſulphuric ſolution expreſsly for that purpoſe, that which has ſerved for the dreſſing may anſwer very well: neither of theſe need be ſtronger than has been there directed, unleſs the goods be conſiderably charged with colour, and there be a great quantity to immerſe at once. The acidulated water is tried by the areometer, and if, in conſequence of having been uſed, it ſhould not be ſufficiently ſtrong, it may be reſtored by adding the requiſite quantity of acid for that purpoſe. It is neceſſary when any new acid is poured in, to mix it well with the water before any goods are immerſed therein.

It muſt, in this place, be obſerved, that though the thread and piece-goods may become charged with a foreign colour, in conſequence of the accidents here pointed out, both theſe articles are frequently very well bleached at the under ſurface. It is even a proof that the muriatic acid has operated effectually, in cauſing the lixivium to produce ſuch an effect: but theſe acidents are difficult to be obſerved on objects ſimply cleared, or in the crude ſtate. In the latter

latter caſe, a permanency of the original colour may alone ſhew the neceſſity of uſing the ſulphuric acid, particularly when the lees and the muriatic acid which have been uſed are not at all exhauſted.

Accidents attending the preparation or dreſſing. When the piece-goods are immerſed in a ſolution of ſoap, after having been taken out of the ſulphuric acid, while they are ſtill too ſtrongly acidulated, or if inſtead of rinſing them they be immediately conveyed from the acid into the ſolution of ſoap, this laſt ſolution is ſubject to curdle, or become immediately decompoſed; whence the operator has the mortification to obſerve the whole ſurface of the goods covered with an infinite number of ſmall ſpots of oil, in the form of clots, of a yellowiſh colour, and very tenacious, particularly on ſtockings or cotton goods, becauſe they incorporate as it were with the nap or texture of the goods: they diſappear in conſequence of much waſhing or rinſing. I muſt particularly mention an accident which may happen to any one, namely, that of placing by miſtake ſtockings or other bleached objects, which have received their firſt treatment in the ſolution of ſoap, upon articles which have been expoſed to the vapour of ſulphur. I have placed ſtockings upon gauze, which had been whitened by by ſulphur, and found that after they had remained in this ſituation for the courſe of a night, they became entirely of a brown-red at the place of contact. They appeared as if burnt or marked with an hot iron. This colour, which, no doubt, was produced by the combination of the volatile ſulphuric acid, with the alkali of the ſoap, with which the ſtockings were ſtill impregnated to a certain degree, immediately diſappeared upon expoſing them, firſt, to the action of a bath of the odorant oxygenated muriatic acid, and afterwards to another of water, ſlightly acidulated with the ſulphuric acid.

Every ſalt with exceſs of acid, ſuch as the ſalt of ſorrel, removes the ruddy ſpots here mentioned with equal eaſe. It is true, that this ſalt cannot with convenience be uſed, on account of its dearneſs, but the reſidue of the diſtilling veſſels, that is to ſay, the water which holds in ſolution the reſidue of the diſtillation of the oxygenated muriatic acid, is very ſerviceable in this proceſs, and may be advantageouſly uſed either hot or cold, to remove thoſe very tenacious ſpots, which are not at all capable of being removed by ſoap or alkaline lees.

CHAP.

CHAP. XVI.

The Method of taking out Spots of Rust or Ironmould, Tar, Fruit, Wine, &c.

WHEN the ſpots of oxyde of iron, commonly diſtinguiſhed by the name of ironmould, are ſmall, they may eaſily be taken out with ſalt of ſorrel in powder, laid upon the ſpot itſelf, which is afterwards to be moiſtened with a ſmall quantity of water; or the part which is ſpotted may be ſteeped in a ſolution of the ſame ſalt. It ſoon becomes fainter, and at length diſappears, after which the place muſt be very well rinſed. The ſulphuric acid may be uſefully applied inſtead of the ſalt of ſorrel, as Bertholet ſeems to affirm in his memoir; and I have proved with ſucceſs, that, though the ſpots may penetrate quite through the cloth, and be very broad, yet if they be ſoaked in a bath of ſulphuric acid, either warm or cold, when the goods are taken out of the bath of muriatic acid, the

the effect will be that the ſpots inſenſibly diſappear. If the goods be of cloſe texture, the operation of the acid is ſlower *.

With regard to the ſpots of ruſt which are frequently ſeen on thread or cotton ſtockings, they are produced by the needles of the engine, and commonly diſappear during the dreſſing, that is to ſay, in the bath of ſulphuric acid. The ſame obſervation is true of the ſpots of ruſt which ſometimes appear on the piece-goods, in conſequence of their having been in contact with iron. In general, the older any ironmould may be, the more tenacious it is, and the more difficult to be effaced; but every ſpot may be made to diſappear in time.

It frequently happens that piece-goods are ſpotted with tar, during their carriage by water, in boats, where they are liable to be placed upon the pitchy parts of the veſſels, or in contact with tarred ropes. Theſe ſpots may be ſoon taken out, by rubbing them with oil of olive, which diſſolves the tar; or ſtill better, by holding the part in ſpirit of wine, if this proceſs ſhould be thought more convenient. The latter

* The ſalt of ſorrel is ſold in London, in ſmall bottles, by the perfumers and apothecaries, under the name of ſalt of lemon. The ſulphuric acid, as preſcribed above, muſt, of courſe, be diluted.—*T.*

method

method operates by the complete ſolution of the tar.

With regard to ſpots of wine, cyder, or any kind of fruit, they may be effaced by dropping a few drops of the oxygenated muriatic acid upon them, which cauſes them almoſt inſtantly to diſappear. But there are certain fruits, ſuch as plumbs, of which the ſpots are more difficult to efface; they requiring one or two lixiviations. Thoſe that are grey, or reddiſh, at firſt, aſſume a fine yellow colour in the muriatic acid, which does not diſappear during a ſubſequent lixiviation, but requires a ſecond immerſion in the bleaching liquor.

I muſt not omit a ſecond very ſimple and economical method to take out every kind of ſpot occaſioned by fruits, ſuch as ſtrawberries, gooſeberries, &c. It conſiſts in cauſing the ſpotted part to imbibe water, and afterwards to burn one or two common brimſtone matches over the place: the ſulphureous gas which is diſcharged ſoon cauſes the ſpot to diſappear.

There is a kind of indelible ſpot which is produced from red ochre and the charcoal black, with which the weavers mark the turns of the beam, in order to aſcertain the length of the chain of piece goods. This kind of mark, which

which is impreſſed on the goods at equal diſtances, is ſo far from being effaced, that it ſeems, in ſome meaſure, to receive ſtrength from the oxygenated muriatic acid, notwithſtanding the intermediate action of the lees.

CHAP.

CHAP. XVII.

The Expence of Bleaching different Kinds of Goods, of Linen, Hemp, or Cotton, by the oxygenated muriatic Acid, at per Ell, or per Pound.

BEFORE I proceed to give an account of the expence of bleaching any quantity of ells or pounds of goods, by the muriatic acid, I ſhall, in the firſt place, mention the prices of the materials required to make the liquor, either with or without ſmell, of which I have before deſcribed the compoſition. The prices are calculated according to thoſe of the articles to be delivered at Abbeville, in 1791 *.

The ſulphuric acid of Rouen, rectified for

* I have not reduced the numbers in this chapter to their values in Engliſh money, becauſe the difference of locality would, even in that caſe, have rendered them of little immediate utility.

As tranſlator, I am obviouſly not at liberty to omit the chapter, even if I were ſo diſpoſed. The Engliſh prices of the materials are given in the Appendix.—*T.*

the market to 66 degrees, delivered at Abbeville, will cost 11 sols the pound, wholesale.

Manganese, crystallized in needles, ready sorted, from Pelletier, apothecary, rue Jacob, at Paris, 8 sols, retail.

Grey muriate of soda, in the market of Abbeville, 2 sols, retail.

Charcoal, weighing five or six pounds to the bushel, according to the quality of the wood, 3 sols, retail, per bushel.

Blue potash, of Dantzick, or the yellow potash, in hard lumps, of York, 12 sols, wholesale.

Green or black soap, of Abbeville, 8 sols, retail.

White marble soap, of Marseilles, bought at Abbeville, 12 sols, retail.

The sack of turf, containing four mannes, 8 sols, wholesale.

I shall now shew the expence of charging two pneumatic vessels, by two successive distillations with the simple apparatus, or by one distillation, when the apparatus has two of these vessels, as well with muriatic acid which emits no smell, as with that in which the odorante gas is not detained.

Expence

Expence of two pneumatic vessels of the muriatic acid, without smell, made according to the proportions prescribed in this Work.

	Liv.	s.	d.
Sulphuric acid, five pounds and a half - - -	3	0	6
Manganese, two pounds and a half - - - -	1	0	0
Grey muriate of soda, eight pounds - - - - -	0	8	0
One bushel of charcoal of wood - - - - - - -	0	3	0
Workman, one day - - - - - - - - - - -	1	0	0
Potash, two pounds and a half - - - - - - -	1	10	0
Total	7	1	6

Expence of two pneumatic vessels of the odorant muriatic acid, made according to the proportions prescribed in this work.

	Liv.	s.	d.
Sulphuric acid, five pounds and a half - -	3	0	6
Manganese, two pounds and a half - - -	1	0	0
Grey muriate of soda, eight pounds - -	0	8	0
One bushel of charcoal of wood - - -	0	3	0
Workman, one day - - - - - -	1	0	0
Potash - - - - - - - - -	0	0	0
Total	5	11	6

I shall, in the next place, shew the expence of lixiviating, or rather of boiling, two pieces of cloth, each seventy-two ells long, of such a degree of fineness, that two ells in length, on a breadth of two-thirds, may weigh one French pound; or of seventy-two pounds of thread,

 such

ſuch as is commonly ſpun in Picardy. I chooſe this kind of cloth in preference, as an example, becauſe it ſeldom happens that coarſer goods are required to be bleached even to the ordinary white, which I muſt be underſtood to mean in the preſent inſtance.

The proportion of ingredients for the lees required to ſteep the ſeventy-two ells of cloth, beforementioned, or the ſeventy-two pounds of thread, are ten veſſels of water, of eighteen pints each, with about five pounds of potaſh, which gives a degree of ſtrength, marking rather more than one and a half, compoſed according to the directions in chap. VII. ſeparately, and in a ſmall covered portable boiler.

Expence of new lees, proper for one boiling of ſeventy-two ells of cloth, or ſeventy-two pounds of thread, of middling quality.

	Liv.	s.
Potaſh, five pounds, 3 livres - - - } Turf, two mannes, or half a ſack, 4 ſols } - -	3	4

It remains to be ſhewn, what may be the expence of bleaching the above ſeventy-two pounds of ſingle thread, or ſeventy-two ells of cloth. I have before obſerved, that cloth of middling fineneſs requires nearly four immerſions, two of which may be made in the muriatic acid without ſmell, and two, if it be thought

thought better, in the odorant acid, beſides four lixiviations; and again, that one pneumatic veſſel is ſufficient for ſixty pounds of thread, at the firſt immerſion, and from ſeventy-two to eighty at the ſecond. I ſhall take ſeventy-two pounds as the middle term, between the firſt and the laſt immerſions, which, as well as the lixiviations, I will ſuppoſe to be made with freſh ſolutions.

	Liv.	s.
Two pneumatic veſſels, for the two firſt new immerſions in muriatic acid without ſmell -	14	3
Two pneumatic veſſels, for the two other immerſions, in the odorant muriatic acid - -	11	3
Four new lixiviations, or the quantity of potaſh neceſſary for that purpoſe - - -	12	0
Two ſacks of turf - - - -	0	16
One day's work - - -	1	0
Total	39	2

Hence, the pound of thread of Picardy, containing ſixteen ounces, will coſt 10 ſols 10 deniers. With regard to Flanders thread, which is cleared in water, the price will not, at moſt, exceed 8 ſols, becauſe this article requires only one immerſion, and a lixiviation leſs. Flanders thread, likewiſe, as has been remarked, is ſubject to a loſs of no more than twenty per cent.

If this calculation be, therefore, applied to the bleaching of common coarſe cloth, two ells of which weigh a pound, the ell will not exceed 5 fols 5 deniers for the bleaching.

If it be, therefore, fettled to charge for the thread of Picardy, or any other which is cleared on the grafs, 12 fous the pound, or livre de Marc; for Flanders thread, or any other, which is rotted or cleared in water, at 10 fols, upon an average; and for linen piece-goods, 8 fols the ell, of fine or middling quality, the manufacturer will find himfelf reafonably paid for his trouble. Thefe are the ufual prices at the moft celebrated bleaching works of Lifle, Beauvais, Saint Quentin, Senlis, Rouen, Rheims, &c. I muft, however, take notice, that the dreffings are not reckoned in thefe charges, which, with regard to the piece-goods, amount to about 2 liards per ell for cold calendering, and 1 fol for hot calendering, including the folding, &c. There are fome articles of which the price of the dreffing amounts to half that of the bleaching: thefe are fuch as require a degree of firmnefs, by means of ftarch, gum, or other fimilar material, with blue, which, in certain markets, and with regard to goods of a certain defcription, is favourable to the fale.

The proper dreffing for thread amounts to about 1 fol the pound, but is the object of a particular agreement between the bleacher and the

the owner. With respect to the price of bleaching double, or sewing, threads, my advice is to charge 2 sols extra per pound, on account of the greater difficulties they present, and the attentions they require, as may be gathered from what has been before said on this subject

Piece-goods, in general, require more care, and are attended with more difficulty, than threads, on account of their volume, their weight, their texture, and the even white colour required to be given, on both sides, as well as towards the selvedges. It must also be remarked, that the selvedges having their texture closed by the action of the temple, when the cloth is in the loom, sometimes require, in the middle of the bleaching process, to be rubbed with soap, by hand, for which purpose black soap is to be used in preference. If they be not opened to the action of the acid by this treatment, there will be danger of the white being less advanced towards the edge than on the rest of the surface. This management may, however, be avoided, if, at the time of steeping and macerating the cloth, as well as in the first lixiviation, the operator is careful to rub or clear these parts, on account of the firmness of their texture.

From these several data, it will be easy to

eſtimate the expence of bleaching finer thread, for linens and lawns, as well as that of the ell of theſe articles reſpectively. For nothing more will be neceſſary, for that purpoſe, than to conſider the account of the number of immerſions and lixiviations which I have ſtated to be neceſſary for thoſe goods. The thread of lawns, of the ordinary fineneſs, will run ſix ells in the piece to the pound, on a breadth of one ell.

Having thus ſhewn the coſt of bleaching thread or linen goods, by the pound or ell, I ſhall proceed to examine that of the ſame articles in cotton; for which purpoſe I ſhall chooſe the thread proper to make the commoneſt wrappers, which run two ells to the pound, on a width of one ell. I have before ſtated, that each pneumatic veſſel is ſufficient for the immerſion of eighty or ninety pounds of thread, for the firſt working, and 100 for the ſecond. I will, therefore, take only ninety for the middle term. I have alſo ſtated, that no more than three immerſions, at moſt, were required to bleach cotton, one of which ſhould be in the muriatic acid without ſmell, and the others in the odorant acid, beſides three lixiviations. I will ſuppoſe that neither the acid nor the lees have been uſed before. My piece of cloth ſhall be aſſumed at 180 ells, or two pieces of ninety

ells

ells each, and the quantity of thread equivalent to this ſhall be ſtated at 90 pounds; whence it will follow:

	Liv.	*s.*	*d.*
Firſt immerſion in new muriatic acid, without ſmell - - - - - - - - - -	7	1	6
Two other new immerſions in the odorant muriatic acid - - - - - -	11	3	0
Three new lixiviations in one of double proportion with regard to the maſs to be lixiviated -	12	0	0
Three ſacks of turf, on account of the double lixiviations - - - - - -	1	4	0
One day's work - - - - - - -	1	0	0
Total	32	8	6

This computation ſettles the pound of cotton thread at about 7 ſols 6 deniers, and conſequently the cloth at 3 ſols 9 deniers the ell. It is to be obſerved, that cotton, being more looſe and ſpongy, and more ſubject to riſe up in the boiler by the action of heat, requires near double the quantity of lees than for thread, and conſequently more fire in the ſame proportion to heat it, ſuppoſing the ſame boiler to be uſed. The ſame remark is, in part, applicable to the muriatic acid; but as this may be uſed rather weaker for cotton than for thread, the liberty of diluting it with water may be taken.

If the bleacher, according to this new method

thod, ſhall therefore fix the price of bleaching cotton thread at 8 ſols 6 deniers the pound, and of cloth entirely of cotton at 6 ſols the ſquare ell, of every kind coarſe or fine, he may derive conſiderable advantage, and the public will have no reaſon to complain, ſince cotton threads in general, and likewiſe muſlins, require much care and attention, on account of the delicacy of their texture and the ſlight tenacity of the fibres, the ſhort ſtaple of which, as is very well known, will ſcarcely permit it to be turned on the reel without great care.

It now remains to be ſhewn what price ought to be fixed for the bleaching of ſtockings of linen or cotton per pair. I ſhall begin with plain thread ſtockings of men's ſize, from which an eſtimate may be made for ſmaller articles of the ſame kind, as well as all other knit or ſtocking-wove goods. I ſhall likewiſe aſſume that one pair of men's ſtockings contains half a pound of thread, and conſequently 6 pounds will be contained in one dozen pair. My calculation will be for 12 dozen or 72 pounds of thread. I ſhall likewiſe aſſume that one pound of green ſoap will be required for the firſt waſhing of ſix dozen pair of men's ſtockings, and one pound of white ſoap for the ſecond and laſt waſhing of the ſame ſix dozen. And accordingly I ſhall add to the ſum before

before deduced, for the mere and ſimple bleaching of 72 pounds of linen, of which the detail has been given, the ſurplus in lixiviations, immerſions, and waſhing with ſoap, which ſtockings require. This amounts to two lixiviations, and the ſame number of immerſions as I have ſhewn at chapter X.

	Liv.	*s.*	*d.*
Therefore firſt ſet down the ſimple price of bleaching 70 pounds of thread, namely - -	39	2	0
To which add two additional lixiviations - -	6	8	0
Two immerſions in the odorant muriatic acid	11	3	0
Two pounds of green ſoap for the firſt waſhing	0	16	0
Two pounds of white ſoap for the ſecond waſhing	1	4	0
Half a ſack of turf to heat the ſolutions of ſoap	0	4	0
A woman one day for the waſhing - -	0	15	0
Total	59	12	0

This account gives 8 ſols 3 deniers for each pair of men's ſtockings; and if 12 ſols be taken for this article, I am of opinion that there will not be many bleachers envious of the advantage of rendering them milk-white at this price, on account of the difficulties they preſent, which require them to be turned from time to time to open the texture, which would otherwiſe become cloſe and impenetrable to the muriatic acid: and if the ſtockings be ribbed, or have clocks, it will not be too much to charge 14 ſols the pair, on account

account of the particular care required for thefe kind of goods, the ribs of which being difpofed to fhrink up, are very apt to prevent the intire action of the acid.

With regard to women's and boys' ftockings, 10 fols per pair may be charged, and for fmaller articles 8 fols; at which laft price gloves ought to be charged, becaufe the fingers being clofer than the other parts, require to be turned from time to time to produce an even colour. Mittens may be charged at 5 fols the pair. Articles of the fame kind of thread and cotton, mixed, deferve nearly the fame price on account of the thread which retards the bleaching.

We muft now inquire the price of bleaching the fame articles in cotton. Here likewife I fhall ground my comparifon on plain ftockings for men, admitting that 6 ounces of cotton will make one pair of this fize, which will amount to 4½ pounds the dozen pair, or 90 pounds of thread for 22 dozen pair. To this laft quantity we fhall direct our inquiries, in which I fhall confine myfelf to add to the former determination with regard to 90 pounds of cotton thread of like quality, the extraordinary lixiviations and immerfions which knit or ftocking-wove articles require. This excefs, as fhown at chapter X. is

half

half a lixiviation and one immerſion in the odorant muriatic acid.

	Liv.	*s.*	*d.*
I ſhall therefore reckon for the mere bleaching of 90 pounds of cotton thread as before ſtated -	32	8	6
To which add half a lixiviation in a double doſe	3	4	0
One immerſion in the odorant muriatic acid -	5	11	6
One ſack of turf for heating the double quantity of lees - - - -	0	8	0
Four pounds of green ſoap for the firſt waſhing	1	12	0
Four pounds of white ſoap for the ſecond and laſt waſhing - - - - - -	2	8	0
One ſack of turf for heating the ſolutions of ſoap	0	8	0
Two days' work of a woman to waſh -	1	10	0
Total	47	10	0

Whence we ſee that the coſt for 1 pair of men's plain cotton ſtockings is about 4 ſols, and if the charge be ſettled at 5 ſols for men and 4 for women, or 4 ſols 6 deniers one with another, there can be no cauſe of complaint. Ribbed ſtockings muſt, however, be excepted. Theſe deſerve at leaſt an addition of 1 ſol per pair for the extra attentions, which have been before mentioned.

Nightcaps may be charged, one with another, at 2 ſols 6 deniers: gloves, on account of the fingers, muſt be charged at 3 ſols the pair, and mittens, and children's ſtockings, at 2 ſols.

The

The foregoing are, in general, the articles which are uſually bleached at the proper works for that purpoſe. With regard to the prices I have aſſigned for bleaching each article without any kind of dreſſing, they are ſuch as I have, from my own experience, thought fit to adviſe to thoſe perſons to whom I have had the pleaſure of teaching this new and important art of bleaching. They are capable of being conſiderably diminiſhed by turning to advantage the lees and acid which have been uſed as I have adviſed in the proper place. But I have choſen in my eſtimates to conſider them as new, in order that I might be ſubject to no reproach for diminiſhing the charges which I have, on the contrary, ſtated at the higheſt, as every operator may convince himſelf. If to theſe firſt ſavings of lees of muriatic acid, and of the other collateral and dependent objects, we add the advantage which may be derived from the old lees, as I have ſhewn, as well as from the exhauſted bleaching liquor, the reſidues of the retorts, &c. there can be no doubt but that all theſe different prices may be conſiderably abated; even though we might not venture to affirm that the expence would be entirely compenſated by the profit ariſing from an intelligent application of theſe matters, which have heretofore been thrown away as uſeleſs.

CHAP.

CHAP. XVIII.

The Method of bleaching yellow Wax, Nankeen Stockings, and other Articles which have acquired a dark Colour by keeping; Linen ſtained by Dampneſs, and the Madder Grounds of printed Goods.

THE bleaching of yellow wax may be effected by means of the bleaching liquor, with ſmell, as well as with that which has no ſmell. For which purpoſe a ſingle immerſion, or, at moſt, two, with the ſame number of intermediate fuſions, are neceſſary. The operation, nevertheleſs, ſucceeds more ſpeedily with the odorant muriatic acid, becauſe the wax bleaches as well above as below; which facility it acquires by its property of ſwimming, and preſenting a greater ſurface, as well to the gas which riſes in the liquor, and bleaches in its paſſage, as to that which eſcapes above the ribbons of wax, which, being retained by the covers of the veſſels, is forced to act upon the ſurface expoſed to its action by falling in a kind of dew. Theſe ribbons of wax muſt be very thin.

It is more convenient, however, to uſe only the vapour of the oxygenated muriatic acid, as as Berthollet informs us from the experiment of Landriani. This laſt method is, as I likewiſe find by experiment, much more effectual. To prove this truth, nothing more is neceſſary to be done than to expoſe wax raſped or ſcraped into very thin leaves, under the cover of a pneumatic veſſel, above the ſurface of the liquor. I am even diſpoſed to think that this experiment may point out the invention of a bleaching proceſs in appropriate veſſels furniſhed with different ſtages of frame-work covered with coarſe cloths. The goods being ſuſpended through the whole height of the veſſel thus conſtructed, or elſe in a chamber diſpoſed and appropriated to the ſame effect, that is to ſay, that it ſhall be provided with ſhelves, or poles, ſo diſpoſed all round within its capacity, upon which the acid vapour, directly conveyed from the neck of the retort, or admitted through the ſides of the chamber, may thus act with great freedom and promptneſs, in the ſame manner as the volatile ſulphureous acid acts upon goods which are required to be bleached by its means. I have alſo remarked, that the maſs of wax, with which the ends of the leaden tubes, plunged in the intermediate veſſel when I uſed this apparatus, was rendered of a beautiful white through its whole thickneſs, which was nearly half a line, and

and this by no more than two hours expoſure to the action of the vapour.

The true nankeen is bleached or deprived of its colour with ſome difficulty. It is firſt to be wetted and wrung; after which it is ſubjected to a firſt immerſion in the bleaching liquor, which deprives it of a large portion of its colour. It is then to be properly rinſed, and agitated in a good ſolution of ſoap, which is preferable to lees, becauſe it cauſes the colour, which had merely diſappeared, to ſhew itſelf again more effectually. The piece of nankeen is then to be well rinſed, and ſubjected to a new immerſion. The number of immerſions varies according to the ſhade of the colour; but this article ſeldom requires more than three immerſions, with intermediate waſhings with ſoap. The finiſh is given in a bath of ſulphuric acid, after which it is to be rinſed in a large quantity of water, and then wrung and dried. This proceſs may be performed indifferently with either of the two acids, the odorant, or that without ſmell; nevertheleſs the latter ought always to be preferred, particularly for the firſt immerſion, becauſe it more ſpeedily and equally deſtroys that kind of fleſh colour which is peculiar to the true nankeen. Nevertheleſs, though the oxygenated muriatic acid acts ſo ſtrongly on this

P colour,

colour, I never have been able to bring nankeens to a white of the ſame beauty as is acquired by cotton, thread, and piece goods bleached by this proceſs.

Stockings and other goods bleached by the old proceſs, but which have acquired a ruddy colour, at the extremity of the folds, by remaining in the ſhop or warehouſe, partly uncovered either from want of care or for ſhew, require only a ſingle immerſion without preliminary ſoaping or lixiviation. The ink marks which retailers are in the habit of making to aſcertain either the number, price, or quality of their goods, partly diſappear in this immerſion, and totally in the bath of ſulphuric acid, in which they are afterwards plunged. Theſe goods are to have all the ſubſequent dreſſings, of which I have given an account, if the proprietor expects or requires it.

It is a peculiar property of the oxygenated muriatic acid, to diſcharge thoſe black ſpots which are ſeen on foul linen, particularly when they riſe from perſpiration or moiſture. The places moſt ſubject to theſe ſtains, are where the linen is applied to the back or beneath the arm pits. This proceſs is to be commenced with a lixiviation or boiling, which is to be ſucceeded by an immerſion, and afterwards by a

bath

bath of ſulphuric acid. However ſtrong the ſpots may be, they never reſiſt theſe ſeveral operations.

Spots of brandy likewiſe diſappear by the ſame proceſs.

With regard to the madder ground of painted or printed goods, it is eaſily diſcharged by either of the oxygenated muriatic acids, a ſingle bath uſually being ſufficient for that purpoſe.

For greater convenience, it is adviſeable to uſe the acid without ſmell, becauſe the operator may, with more eaſe, follow and conclude at a proper time the immerſion of the piece as ſoon as, while paſſing it over the reel, he obſerves that the ground is ſufficiently white and clear. The bleaching liquor, which is partly exhauſted, may be uſed to advantage in this proceſs. Before the immerſion is made, care muſt be taken to plunge the piece in water and wring it out ſo far as to leave it merely humid or moiſt. After the immerſion it muſt be well rinced and dried either in the ſun or in the ſhade, turning the coloured ſurface from the ſun.

It may be remarked that the deep reds are capable of being again brought out, or rendered ſlightly red, by the ſun's light, and the other ſhades advanced in proportion; this would happen in the common method of bleaching, if the printed

 part

part were not always turned to the graſs. I ſhall proceed to mention ſome circumſtances with reſpect to this method of diſcolouring or bleaching, which may be of uſe to thoſe who are intereſted in applying it to practice.

Goods printed in faſt colours *(bon teint)*, for thoſe with chemical colours *(petit teint)* are too difficult to be treated by this method, intended to be bleached by the oxygenated muriatic acid, inſtead of the uſual expoſure in the field, ought to have their deſigns much more charged with colour, than ſuch as are intended ſimply to be ſubjected to the action of the air; in order that while the acid exerciſes its action on that part of the ground which is maddered without mordant, the ſame action which is alſo exerted on the part where the madder is combined with the mordant, may not deſtroy in the laſt part any more of the colour than that quantity in exceſs, beyond what the piece ought to preſerve to produce the intended effect, and, conſequently, that it ſhould not, after the proceſs, appear more ſenſibly altered than it would have been after the uſual expoſure in the field: this precaution ought to be attended to more particularly with regard to the ordinary violets, blacks, and browns. They are much more eaſily degraded than the red, or roſe-colours, and the deep browns.

One

One leading object, which is essential to the preservation of the colour, and contributes infinitely to the unmaddering, is to give the pieces one or two boilings in bran and water, which may follow a boiling in a solution of soap. If these three boilings are properly managed, the ground of the piece goods will be brightened at least three quarters. One or two immersions in the bleaching liquor will remove the small portion of colour which remains. Between the two immersions attention must be paid to plunge the goods in bran and water. This ought to be done after the last immersion, for it raises and relieves the tone of the colours which may have been slightly weakened.

I have several times found, that when the preparatory baths have been well proportioned to the grounds intended to be coloured, it is unnecessary to apply the bleaching liquor. A few days exposure in the field are sufficient afterwards to complete the bleaching.

The proportions which I followed for the bath of bran and water, were three ounces and an half of wheat bran, and three pounds and an half of river water. Those for the solution of soap were two ounces of soap to four or five pounds of water; the weight of the goods to which these doses were adopted were 10 gros.

 More,

Moreover, it is practicable, according to the depth of the tints, and the experience the operator may have acquired, to diminish the force of the oxygenated liquor to that point which may insure him against a too perceptible destruction of those parts which ought to preserve their brightness. But, in this practice, the process is too slow, and the great advantage of using this method with regard to such kind of goods would thus be lost.

In a considerable manufactory, where the operations of printing and bleaching succeed each other with rapidity, it might, perhaps, be more advantageous to clear off no more than three-fourths of the ground of the cloth which has received the madder without mordant, by subjecting it to the boilings with bran, water, soap, and a slight immersion in the liquor, as has been prescribed, and afterwards to expose it to the action of the air in the field. This method of operating seems preferable, and would be no less expeditious, whether in summer or in winter. And in the case only of the goods being in great haste, the complete bleaching with the oxygenated acid might be adviseable, taking care to use all the precautions which have already been pointed out.

In order to avoid exposing the goods to too con-

conſiderable alteration, it is more convenient to paſs only one or two pieces through at a time, with the attention that they ſhould be of the ſame degree of intenſity in their colours, in order that if it ſhould be neceſſary to ſtop ſuddenly the effect of the liquor, it may, in ſome meaſure, be done inſtantly. This could not be eaſily accompliſhed, if eight or ten pieces were ſewed together as ſoon as ſoaped. For it is eaſy to imagine, that while one piece was drawn out, the others remaining in the liquor too long a time, would be expoſed to have their colours weakened, at leaſt in the proportion of the longer time they remained in the liquor. If it were thought an object of ſufficient importance, this laſt inconvenience, however, might be removed, by placing at the bottom of the veſſel for immerſion a platform of ſtrong baſket work, which might be ſpeedily raiſed by means of a pulley, or other mechaniſm, to remove the goods out of the bath at the inſtant it might be found neceſſary, and they might afterwards be thrown into a reſervoir of water, or conveyed to the river.

It would be a very deſirable object, if the oxygenated muriatic acid could act only upon the ſurface oppoſite to the printed ſide of the piece. In this way it would operate like the atmoſpheric air, without giving cauſe to fear the deſtruction or perceptible alteration of the ſhades, whatever might be their depth. The difficulty of ſucceeding, and the length of time required for bleaching, may, perhaps, ſooner or later, give riſe to a method of fixing the colour by particular mordants, without the aſſiſtance of madder.

It would alſo be a deſirable object to diſcover a proceſs to prevent too much degradation of the tranſverſal red or blue ſtripes, and other ornaments of coloured thread, which are uſually made at each extremity or angle of cotton coverlids. This proceſs might alſo be applicable to the defence of thoſe tranſverſal blue or red ſtripes which are made in pieces intended for napkins, between one napkin and another, and at the ends of pieces of muſlins and the like. The beſt method, no doubt, would conſiſt in manufacturing theſe goods entirely of one colour, and afterwards making the terminations with coloured thread. As theſe kind of goods are capable of being ſoiled, either by the dreſſing given to their chain, or by the different operations which ſucceed or are previous to the weaving, they may be eaſily cleanſed by a proper waſhing or ſoaping.

The following is the expedient which I have thought

thought proper to uſe to preſerve the ſtripes in queſtion from every action which might be too perceptible. After two good baths in the lixivium, each of the ſtripes was covered, on both ſides, with one or more coatings of chalk and oil, which was left to dry until the pieces could be handled without fear of ſpotting the neighbouring parts. I then ſubjected them to the bleaching liquor, afterwards to a ſlight lixiviation, and a ſolution of ſoap, &c. and ſo on, ſucceſſively, till the ground was as clear as required. After each lixiviation I took care to repair or renew the covering, if neceſſary. I afterwards cleared off this covering of chalk and oil, either with a good ſoaping, or with a ſlight bath of ſulphuric acid, according to the nature of the colour of the bar, and the degree of tenacity of the paint. If by accident the colour of the bar was ſomewhat weakened, it did not fail to be raiſed again, by paſſing it through bran-water at the concluſion of the proceſs. I apprehend that this expedient, which I have always uſed with a certain degree of ſucceſs, will be acceptable to the manufacturer.

CHAP.

CHAP. XIX.

The Method of discharging the Colour of painted or printed Cottons, or Linens, and every Kind of Dye on Cloth or Thread, before or after it is wrought up.

ALL the colours of callicoes, or printed goods in faſt colours, are deſtroyed by either of the oxygenated muriatic acids, without having recourſe to the lixiviations or other previous or intermediate operations before deſcribed. The blues, yellows, and blacks, afford an exception with reſpect to the bath of ſulphuric acid, which muſt be ſubſtituted inſtead of the lixiviation. A ſingle immerſion in the muriatic acid is ſufficient to deſtroy all other colours, ſuch as reds, yellows, auroras, green, &c.; but the yellows, properly ſo called, and the lemon colour, with which greens are produced, and the blues and blacks, ſometimes require, according to their ſhade, three immerſions, and two or three intermediate baths of ſulphuric acid.

It

It muſt not, however, be ſuppoſed, that the Adrianople reds, when diſcharged by the oxygenated muriatic acid, become perfectly white. There always remains a ſlight ruddy appearance, which ariſes from the oily matter which enters into the preparation for this dye.. This tinge of redneſs does not diſappear, however numerous the lixiviations and immerſions and baths of ſulphuric acid may be.

There is another thing no leſs worthy of remark with regard to the black colour, which forms the outline or border of deſigns, namely, that if the muſlin, or cleared fine piece, upon which the different flowers were deſigned which have been diſcharged, be folded together in ſeveral folds, or placed upon a dark-coloured ground, the effaced outline becomes viſible according to the expoſure of the piece under a certain obliquity of the light exhibiting the appearance of a ſlight trace. The kind of outline which, under theſe circumſtances, becomes viſible, cannot be compared to any thing better than the embroidery of muſlins placed on a coloured ground. This trace ſeen at a certain diſtance has the ſame effect, and even when cloſely obſerved, it is impoſſible to determine what it is, becauſe it is not viſible, except under a certain reflection of the light; nevertheleſs

theless the whole piece appears white, and of a very superior quality. I have remarked that this effect does not take place excepting with regard to the old prints of flowered designs of the true India callicoes imported from that part of the globe. For in the printed goods of our manufactures, such as those of Paris, Joly, St. Denis, and Beauvais, all the traces of the designs completely disappeared, to my great surprize. It must, therefore, necessarily be admitted, that the difference in these results depend on the qualities of the mordants, which are more or less oily, or the manner of striking the blocks in the act of printing.

If this effect were produced by the mordant with the outlines of the designs in the pieces of printed goods, it might, perhaps, be of advantage to take the same method of obtaining a substitute, instead of the rich expensive embroideries with which the fine muslins of India and Switzerland are covered. These designs likewise do not appear in their full effect, but when they are placed upon a transparent stuff of a deep colour, which exhibits all the outline. This method of producing so rich an effect would be extremely simple, singularly permanent, and highly economical. I think, however, that I may add, that, after many trials,

I have

I have at laſt ſucceeded in diſcharging this mordant, ſometimes by a bath of ſulphuric acid, rather ſtronger than uſual, and at other times by ſoaping the goods before and after the bath. This management is very eſſential to be known, in order that the operator may not be expoſed to the mortification of ſeeing the ſame deſigns return again by the ſecond action of the madder applied to the ſame bleached piece in a ſubſequent printing proceſs. To obviate every accident of this kind, it will be proper to inform the owner which of the methods have been uſed to bleach their goods, and in caſe the new method may have been uſed, it would then be prudent to paſs them previouſly through a good bath of ſulphuric acid.

With regard to chemical colours, as they are called, which are applied on callicoes or other goods, they diſappear immediately, and much more ſpeedily, than faſt colours. A ſingle immerſion in the weakeſt oxygenated muriatic acid, without any other preparation, is ſufficient to deſtroy them, excepting only the outline of the flowers, which, as has already been remarked, requires particular precautions.

Among the yellow colours of this deſcription there is one, however, in the compoſition of which ſulphate of copper, ſulphate of iron, and

and acetate of lead, are ufed, which is fo far from being deftroyed by the oxygenated muriatic acid, that, on the contrary, it is fixed by that means. This colour cannot be difcharged, unlefs the piece be previoufly well rubbed in a good bath of foap, which difpofes it fo far to detach itfelf from the goods, that the immerfion it afterwards undergoes in the muriatic acid completes its difcharge.

It is very remarkable, that, after the difcolouring of the printed goods, particularly brown or black, and when the oxygenated muriatic acid has combined with the merchandize, there arifes from the trough a certain gas, which acts upon and irritates the organ of fight only, to fuch a degree, that it is very difficult to fupport its action for any confiderable time without a difcharge of tears *. This effect, however, is not very perceptible in a trough over which the workman has operated for the whole day, until towards the evening, whence it fol-

* The effect of this gas upon the human body is very fimilar to that which was produced on the 22d Brumaire in the evening, the prefent year, 6th of the republic, by a thick mift, of which the influence was felt, more or lefs, throughout Paris. This action was exerted principally by an irritation of the throat, a pricking fenfation in the eyes and nofe, and a difcharge from the head. The oxygenated muriatic gas produces the fame effects when it is breathed for any length of time, on which fubject fee chap. vi. of the prefent work.

lows

lows that the acid does not act till after a considerable time upon the mordants, so as to produce this peculiar gaseous combination, which is then capable, by its quantity, of irritating the organs of sight. These inconveniences may be avoided, by taking care to work these pieces under a glass cover, expressly disposed for that purpose, nearly as is represented in fig. 1 and 2. in plate 1; or by making use of the covered vessel represented in fig. 1 and 2. plate 9. It would be an important acquisition to know the nature of the gas here spoken of.

With regard to piece goods dyed before or after the weaving, whether of thread or cotton, all the false dyes, such as red, blue, green, flesh-color, orange, grey, black, &c. disappear in an instant, and almost constantly, by a single immersion, and certainly by one immersion and one lixiviation; but it is most usual to omit the lixiviation.

The same remark does not apply to the true dyes, or fast colors, such as blues, Indian red, strawberry colour, deep brown, &c. the yellow colour and lemon colour either applied to linen or cotton: these are much more difficultly effaced. They sometimes require one lixiviation between two immersions, according to the force of the shade. The blue in particular is the most tena-

cious

cious colour; it muſt be obſerved, that a bath of ſulphuric acid muſt always be given at the concluſion, particularly with reſpect to the yellows, of the colour of ruſt of iron, which does not totally diſappear but in this laſt fluid.

With regard to goods which have not been maddered, and of which the deſigns have been printed in oil, the firſt preparation is that of the lees, in which they muſt be heated, without rinſing or clearing off. After this, while they are yet hot from the lees, they muſt be ſtrongly rubbed in a good ſolution of ſoap. Moſt of the colours are, by this means, partly diſcharged, and their deſtruction may be completed, either by the oxygenated muriatic acid, or by the ſulphuric acid. It is ſeldom neceſſary to repeat this courſe of operations, many of theſe colours being uſually diſcharged by the ſoap.

It is certainly proper to remark, in this place, that the effect of the oxygenated muriatic acid in deſtroying all colours whatever, as well on printed goods, as in ſuch as have undergone the proceſs of dying, whether in the thread, or in the web, muſt afford many perſons the advantage of multiplying, in ſome meaſure, the changes of their clothes, without going to the expence of new: for if the old or unfaſhionable colours of a garment be diſcharged, and it be afterwards ſent

to

to the printer's to receive a new deſign, this ſimple proceſs would enable the wearers to change the faſhion every ſeaſon, if they thought proper. The only expence would be that of diſcharging the old colours and printing new, at ſo much an ell, for the ſeveral colours, according to their reſpective value. It is equally obvious, that dealers in printed goods * might, by this means, vary or enlarge their ſpeculations. I muſt likewiſe add, it might be poſſible to take advantage of the ſaid property of the oxygenated muriatic acid, to deſtroy the colours of dyed goods, or to trace any required deſign with the pencil, the pen, ſimply in the way of outline, and in the manner of goods printed *à la reſerve.* I have ſeveral times attempted to ſketch different ſlight deſigns on ſuch goods, principally in the muriatic acid without ſmell, and I ſucceeded perfectly in obtaining very neat and fine traces. It might be, perhaps, an object of ſtill greater intereſt, to give a roſe or other colour to piece-goods dyed *à la reſerve,* becauſe this method has not hitherto been applied but with reſpect to blues, and

* It ſeems probable that the wholeſale method of operating in England, and the effect of the exciſe laws, would render the practice here deſcribed not very convenient in the Engliſh market.—*T.*

ſometimes to orange or olive colour, or a few other light colours of this kind.

The ſame thing might be done with regard to the particular deſigns or things which might be imitated on ſtriped goods, the threads of which were dyed before the weaving; from which it might be poſſible, either to take away part of the colour, or to add at pleaſure a ſtripe of another kind. I have ſometimes accompliſhed this purpoſe on blue and white ſtockings, as well as ſtriped and chequed piece-goods, by lightly ſprinkling the oxygenated muriatic acid upon them: the different drops of the acid produced a ſingular effect by ſpotting thoſe ſtockings and ſtriped goods. All the goods thus treated may be waſhed with ſoap or lees, without danger of effacing the different ſingular marks or deſigns which have been traced upon them. The obſervation I have made, with reſpect to dyed goods, is likewiſe applicable with regard to certain patterns of one uniform colour; it is probable that the prints might be made from a block impregnated with the muriatic acid, combined or amalgamated in ſuch a manner as to work with the ſame accuracy as in the common practice of callico printing. I ſhall hereafter relate an experiment

periment which I have made in ſupport of this obſervation.

I muſt not omit the remark, that the oxygenated acid is very uſeful to brighten up the white deſigns reſerved in piece-goods printed *à la reſerve.*

It is well known that theſe white ſpaces are ſeldom clear; either becauſe the compoſition being ill applied, or ill made, ſuffers a ſmall quantity of the colour to paſs through; or from the effect of the ſulphuric acid in which they are ſteeped to clear off the compoſition, when it is made of tobacco-pipe clay, &c. If the piece, when taken out of the boiler, is not well cleared of its colour, this laſt will ſlightly extend itſelf towards the blue, which is uncovered; conſequently, by ſteeping the piece in a bath of oxygenated muriatic acid, after its immerſion in the ſulphuric acid, the colour is not only brightened, but the reſerved white, in conſequence of the neat finiſh in its outline, is rendered much more ſtriking in its effect.

CHAP. XX.

The Methods of taking out the Dye from Silk and Wool.

SILKS dyed in a ſimple colour, ſuch as indigo-blue, lilac, crimſon, and grey, are capable of loſing their colour, and acquiring a yellow chamois colour, by ſteeping in a bath of oxygenated muriatic acid, without any previous or intermediate lixiviation or preparation. White ſilk receives the ſame yellow colour, if expoſed to this acid. But it is poſſible to convert this yellow colour to white, by expoſing the ſilk to the vapour of ſulphur or the ſulphureous volatile acid. For this purpoſe, it is neceſſary that they ſhould be yet in a moiſt ſtate, to facilitate the equal action of the ſulphureous gas. It muſt be obſerved alſo, that the goods ought not be expoſed too near the flame of the ſulphur, becauſe the heat dries them, and retards the action of the

the volatile acid, and may likewiſe give them a ſcorched or brown colour.

Compound colours, ſuch as browns, violets, greens, and blacks, likewiſe loſe their colour, and acquire a ſimilar tinge of chamois yellow; but this diſcolouring commonly requires two immerſions. The blue of brown violet and puce colours commonly diſappears firſt, leaving the ſhade of red more or leſs weakened. The ſame gradation takes place with regard to the green and orange colours, of which the yellow gives way firſt. The blue of the former, and the red of the latter, only remains. It is neceſſary that the oxygenated muriatic acid ſhould be weak *(légère)* otherwiſe it would acquire an aurora colour inſtead of a roſe colour, when it afterwards came to be ſteeped in ſulphuric acid; for it is to be noted, that it is proper to uſe a bath of ſulphuric acid, and rinſe off with much water previous to each of the ſaid immerſions. With regard to black ſilks, the brown diſappears firſt, and leaves the blue ground, if this may have been uſed; or the root ground, ſuppoſing this laſt to have been the baſis of the black.

Theſe obſervations, reſpecting ſilk, hold good alſo with regard to wool dyed grey, orange, green,

Saxon blue, apple green, root or fawn colour, brown lemon, and dipped blue. All these colours disappear more or less readily, and become of a chamois yellow, like the silk; but this last tint is easily brought to the original white, by exposure to the volatile sulphureous acid. Two immersions in the oxygenated muriatic acid are sometimes required, according to the depth of the colour; and the exposures to sulphuric acid will likewise require to be occasionally repeated. For if the chamois colour does not totally disappear at the first exposure, it will at the second.

If we attend, for a moment, to the alterations which woollen and silken goods undergo by exposure to the air, we shall immediately see, that the oxygen of the atmosphere is the principle which acts on the colouring matters with which the goods are impregnated, particularly the false dyes; and that the change is of the same nature as that which is produced by immersing these goods in a liquid which is, in some measure, saturated with that principle. The difference consists only in the speed with which this effect is brought about in the latter case.

The yellowish colour produced by the oxygen of the air is particularly observable in grey woollen with a raised nap, and loose hosiery of the

the ſame colour. This mode of fabrication and openneſs of texture probably affords a ſtronger hold to the oxygen, from the more extended ſurface it preſents.

With regard to dyed ſilks, thoſe of a roſe colour, and Saxon blue, as well as the falſe blacks, are moſt ſubject to alteration by ſimple expoſure to the air.

CHAP. XXI.

Dying by the oxygenated muriatic Acid.

I HAVE little to ſay, with regard to the dyes, in which the concurrence of the oxygenated muriatic acid is of advantage, after thoſe of the nankeen and lemon yellow, of which I had occaſion to ſpeak in the fifteenth chapter. I ſhall here ſpeak only of the different tinges of grey, which are obtained by plunging white wool or ſilk in a ſolution of ſulphate of copper, and afterwards taking them out and immerſing them in a bath of the oxygenated muriatic acid, either with or without ſmell. By this treatment the operator will obſerve the gradual appearance of a fine grey colour, more or leſs dark, and varying in its tinge, accordingly as the ſolution of the ſulplate, or of the oxygenated acid, may have been concentrated. This dye appeared to me to be ſolid; for I perceived no alteration in its ſhade after expoſing it for ſeveral days to the ſun, and to a ſtrong ſolution of ſoap.

It may be proper, in this place, to ſpeak of a black or grey dye varying in its ſhade, which I have ſeveral times ſeen ſucceſsfully made, in thoſe thofe glafs-houfes where the mineral alcali, with the crude foda, or the purified falt, is ufed. I here fpeak only of the Spanifh foda, which is well known to contain a certain quantity of the muriate of foda, the prefence of which is indicated at the moment of the fufion of the glafs: for at this period more efpecially, as well as during the whitening of the frit, there exhales from the pots, for about half an hour, a thick white fume of muriatic acid, which acts on the nofes and mouths of the workmen, and caufes them to cough and fneeze: its prefence is likewife manifefted by the ruft which immediately covers the pipes and other iron implements, placed within its reach, which the workmen are obliged to brighten, whenever they ufe them. I have concluded that the dye, of which I fhall here give a concife account, was the more evidently produced by the action of the oxygenated muriatic acid, becaufe a confiderable quantity of mangancfe is ufed in the glafs-works in which it is practifed. The quantity ufed is fuch that the *cadmia fornicorum* of thefe works are of a perfect violet colour.

The following is the procefs I have feen in practice, at the glafs-works of St. Gobin, in the department of Aifne, as well as in that of Tourla-ville, in the department of la Manche. The former

former of theſe eſtabliſhments uſe wood for fuel, and the purified ſalt of the ſoda of Alicant; the other burns pit-coal, and uſes the ſame ſoda in the crude ſtate.

The ſkeins of thread being previouſly waſhed in lees, or cleaned and afterwards rinſed and dried, are ſteeped in a ſolution of alum, in river-water. When they are well ſoaked in this ſolution, they are daſhed ſtrongly upon a kind of blackiſh ſoot, which is fixed along with the ſaline vapours by the internal projection of the furnace above the glaſs pot or crucible. After having repeated this a number of times, in order that the thread may become more or leſs loaded with the ſoot, it is agitated or rinſed in the ſame alum water, and again daſhed againſt the ſoot, until it is thought to have acquired a colour ſufficiently equal and deep. Laſt of all, they are rinſed in the ſame water, in which they become deprived of the exceſs of ſaline and colouring matter; after which they are ſlightly rung out, and dried, either in the ſun, or in the ſhade. This black or grey colour, which the thread has thus acquired, is ſingularly tenacious. I have ſtockings of thread, thus dyed ten or twelve years ago, which have been waſhed in lees upwards of forty times, and have loſt not the leaſt portion of the intenſity of their colour. It is to be remarked; that

that linen and cotton piece-goods are dyed by the ſame proceſs. There is no doubt but it would be poſſible to imitate this dye, with profit and advantage, by ſome direct manufacturing proceſs. I have made ſome trials which have ſucceeded, to a certain degree, by putting the ſoot of pit-coal into alum-water, in which I ſteeped thread, which acquired a ſhade, and was afterwards put into a bath of oxygenated muriatic acid. I repeated this alternation ſeveral times, which appeared to communicate an equal dye, and this dye was very ſlightly altered by ſoap.

We may likewiſe give the ſame grey or black ſhade to cotton, by boiling it for ſome time in a certain quantity of the ſaline ſoot of the glaſs-works, uſually diffuſed in water, in which mixture the thread is ſimply turned and worked for a number of times, without any previous or ſubſequent operation, excepting that of waſhing or rinſing, which is always indiſpenſable. I have, in this manner, dyed white thread ſtockings of a violet grey colour *(griſdelin)*. This ſhade became ſomewhat pale after repeated waſhings.

CHAP.

CHAP. XXII.

Various Properties of the oxygenated muriatic Acid.

THE power of difcharging every kind of colour from painted or printed goods, muft render the difcovery of the oxygenated muriatic acid of the higheft value to manufacturers of paper, who may very profitably avail themfelves of the acid to form white paper out of coloured rags. It, in fome meafure, affords them an additional refource to fupply their manufactories with raw materials, and to avoid any particular forting. They may, even in this refpect, extend their fpeculations to cordage, oakum, old fails, and other articles, which they may bleach as fpeedily, and in as large a quantity, as they pleafe, without giving themfelves any concern about the fcarcity of rags. It may alfo be queftioned, why the bleaching property of the oxygenated muriatic acid fhould not be ufed to whiten paper which has been written

written upon, and is become waſte. This paper may afterwards be ſized again, like any other ſort, by which means the product and activity of this manufactory may be inſtantaneouſly augmented. This laſt object is ſo much the more eaſy to be attained, becauſe the leaves of paper, containing writing, require to be ſteeped only one ſingle time in the oxygenated muriatic acid without ſmell. The work is, therefore, of the greateſt facility. This firſt operation may be made on a number of leaves together, diſpoſed in ſuch a manner that the oxygenated muriatic acid may ſurround and penetrate each leaf ſuſpended in the fluid. It muſt be followed by a bath of ſulphuric acid, of the ſame ſtrength as has already been preſcribed for the dreſſings. This bath is eſſentially neceſſary, however clearly the ink may appear to have been diſcharged when the paper comes out of the muriatic acid. The ſulphuric acid is required to take up the iron, which, as is well known, compoſes a great part of every writing ink. Care muſt be taken to waſh the paper, when it comes out of this laſt bath, in clean and limpid water, in order to carry off the ſulphuric acid, after which the paper may be ſized, if neceſſary, and then left to dry. Such paper as has been ſized before it has undergone this operation,

ration, will not require ſizing again, or at leaſt this is very ſeldom the caſe, unleſs it has remained too long in the rinſing water. The paper, when dried, muſt be afterwards treated exactly in the ſame manner as if it had been newly manufactured. This method of bleaching written paper may alſo be of the greateſt uſe to men of buſineſs of every deſcription, merchants, and others, who uſe many books. When theſe have become uſeleſs, and out of date, they may, by the method here directed, be eaſily cleared of their writing, and rendered uſeful a ſecond time *. When we reflect on the property of the oxygenated muriatic acid to diſcharge ink from paper, we obſerve, in the action of this liquor, a kind of analogy with that action which takes place, in the courſe of time, with reſpect to ancient writings. There is reaſon to think, that, in this laſt caſe, the air, by virtue of the oxygen which it contains, is acted upon in the ſame manner as the oxygenated muriatic acid; for old writings are

* Since theſe experiments I have had occaſion to make others, as well on the bleaching of the paſte of paper, as on diſcharging the colour of written or printed papers. I have, therefore, thought it uſeful to inſert, at the end of this work, the ſeries of particular proceſſes which I have made uſe of, and which I addreſſed to the different committees of the National Convention, in the year II. of the republic.

are ſo conſiderably altered, that a ſingle ſlight bath of ſulphuric acid is often ſufficient to diſcharge them entirely, and in caſe this bath ſhould not have been ſufficient, they do not reſiſt a very ſlight immerſion in the muriatic acid.

The ſame obſervation may be made with regard to ſnow and dew; both theſe ſubſtances diſcolour and ſoon render the ſoles of ſhoes yellow when expoſed to their action. This obſervation may very eaſily be made after walking out on the ſnow or graſs embibed with dew.

On the ſame principle it is in the mountainous country, in the department of La Somme, the country people clear their linens ſimply by expoſing them in the winter to the action of the ſnow, the dew, and the miſts, without giving them any other preparation, except that they are careful to turn them from time to time on the ground, for about fifteen or twenty days, during which time this vegetable ſubſtance is expoſed to the influence of the air and the atmoſphere.

A ſolution of ſulphate of ſoda, and the reſidue of the diſtilling veſſels, is ſometimes ſufficient to diſcharge theſe ancient writings, which are already in part effaced by the oxygen of the atmoſpheric air in the courſe of time.

I muſt here mention an obſervation I have had

had occaſion to make in the courſe of my operations on the bleaching of threads and cloths. The waters which had ſerved to rinſe the ſingle, double, and twiſted threads, when taken out of the lees, were very ſoon covered in the veſſels, where this rinſing was then performed with a kind of lather, more or leſs white, accordingly as the thread was more or leſs advanced in its bleaching. This froth, which roſe to the top of the water, was ſometimes more than an inch in thickneſs, according to the quantity of thread or cloth which was rinſed or cleared, and forms an excellent paſte for the immediate manufacture of paper.... It may, in fact, be eaſily underſtood that this ſubſtance is of the ſame nature to that which is uſually formed by the decompoſed rags in the paper-mills, and is, in the preſent caſe, formed of an aſſemblage of the filaments of thread or cloth detached by the lees or the acid, but more particularly by the former, and more ſpeedily and effectually ſeparated by the rinſing. I can alſo aſſert, that the ſamples of paper which I have attempted to make with this material, were very beautiful and fine. The bleachers may, therefore, reſerve this produce, and ſell it to the paper manufacturer at a price which muſt neceſſarily vary according to its colour and quality.

The ſame remark is applicable to the inner part

part of the ſides of the veſſels or tubes which is directly bleached by the action of the oxygenated acid, which, in proceſs of time, renders it of a very fine white colour. This ligneous ſubſtance, when collected, is alſo very proper to form paper, after it has undergone the previous action of the mallets or cylinders of the paper-mill, and is afterwards properly diluted with water, according to the practice of the paper-makers; a very conſiderable quantity of this paſte may even be collected in a ſhort time. Nothing is more neceſſary for this object than to diſpoſe the wood in the veſſels deſtined for this purpoſe in ſuch a manner, that it may preſent alternately to the acid and the ſalt of the intermediate lees the greateſt quantity of ſurface poſſible. Two lixiviations, and two immerſions, are ſufficient to alter the wood ſo far that it may be raſped off with advantage. This very economical method may, with much profit, be uſed to ſupply certain paſtes, which will afford very fine and good paper, according to its beauty, its whiteneſs, and the proportionate mixture of other paſtes formed from rags. I have in this manner fabricated ſmall ſamples of paper, which I ſhewed, at the beginning of 1789, to the adminiſtration of commerce, announcing this particular method, as well as that of making a kind of grey paper with the tufts of the typha paluſtris. Theſe paſtes are not to be rejected, even ſuppoſing they could only be uſed for the white-brown or common paper, or for paſteboard; as they would always contribute to render the fine rags more abundant for the manufacture of white papers, to which uſe they might be entirely reſerved, if it ſhould not be found advantageous to mix them with the other materials to produce the intermediate kinds of paper.

Olive oil, expoſed in the uppermoſt falſe bottom of the pneumatic veſſel to the gas or vapour of the oxygenated muriatic acid, paſſed through water loaded with potaſh in the proportion I have pointed out, became changed to the conſiſtence of ſoft ſoap, or very white greaſe, without taſte, nearly miſcible with water, not ſoluble in the ſpirit of wine, nor ſubject to any perceptible change by the ordinary muriatic or nitric acids. Rectified ſulphuric acid alone decompoſes it almoſt as ſoon as poured on. The muriatic acid, with which the oil was combined, flies off, and the ſulphuric acid changes the white and ſoapy colour of the oil into a brown maſs, which very ſoon afterwards became blackiſh. Does not this experiment lead to a preſumption that it might be

poſſible

possible to form a kind of muriatic soap in the soft or hard form, which should have the property of bleaching? Thus much is certain, that from this notion I have attempted to combine olive oil with potash, partly neutralized by the oxygenated acid; and a sample of thread, which I bleached and soaped with this kind of soap, appeared to me to become white to a higher degree, and much more speedily, than by the method above described. This new method would be of infinite utility in every respect.

Copper or brass exposed in the same manner as the oil above mentioned to the action of the oxygenated acid gas, became, at first, blackish, after which it was covered with a firm, dry, pellicle of verdigrease, as well above as beneath: this verdigrease was of a very fine colour. When washed and ground, it is absolutely equal in colour to that fine English green so highly esteemed, with which the fashionable paper-hangings are printed. It might be possible to obtain this matter, in great quantities of it, at a low price, by constructing an apparatus for this purpose. I have obtained this kind of verdigrease by putting copperplates into the waters obtained from the residue of the distilling vessels. I have had occasion to remark on this subject, that

that the fluid was, in the courſe of time, covered with a pellicle ſimilar to that which riſes upon milk when ſet to boil, but of a green colour.

Water, impregnated with the gas, has no action upon the copper, except in the courſe of a long time; but the gas itſelf acts inſtantly either upon copper or braſs.

This kind of verdigreaſe may alſo be uſefully employed in dying, and, in many inſtances, ſupply the place of that which is made with the refuſe of grapes in the ſouthern provinces.

Tin veſſels (pewter) are totally diſſolved or corroded by the oxygenated muriatic gas, and aſſume a grey colour.

Malacca tin is corroded in like manner, but it aſſumes a whitiſh colour. From this experiment it is that we have thought proper to conclude, that the ſolder of leaden tubes cannot long reſiſt the action of the gas or liquor which is impregnated with our acid, and that it is particularly neceſſary, when tubes of this metal are to be uſed, that they ſhould be caſt entire, or without ſolder *.

* As the muriatic acid, whether oxygenated or not, when in the expanded or vapourous ſtate, attacks, and ſpeedily ruſts copper, iron, and tin, it is improper to have in the place of diſtillation, any veſſel or inſtrument made of thoſe metals, becauſe they would ſpeedily be deſtroyed.

Sheet

Sheet lead did not all, or ſcarcely at all, change its colour or properties by this expoſure. It merely acquired a ſlight brown tinge. It was in conſequence of this experiment that I determined to ſubſtitute tubes and adopters of lead inſtead of thoſe of glaſs, and to recommend that the pneumatic veſſels to be made of common wood, and that theſe, as well as the veſſels for immerſion, ſhould be defended with ſheet lead.

Litharge of gold, or yellow litharge, remains dry, and undergoes no other effect than to acquire a violet colour.

The directions or addreſs on the outſide of letters diſappear entirely, without leaving any trace or alteration in the paper. This experiment, added to that of taking out the ink-marks made by the proprietors of ſtockings, gave me the firſt hint to apply this method of bleaching to written paper, which I have mentioned in this chapter.

Red ſealing-wax became of a pale roſe-colour, and was reduced into a kind of moiſt or ſoft wax.

Indigo, in ſmall fragments expoſed in the ſame manner to the oxygenated muriatic acid gas, changed its colour from a deep blue to the yellow colour of dead leaves. Black pitch merely became red at its ſurface. Hair, and feathers of a black colour, were changed, the firft grey, and the latter to an aurora colour. Green oil-cloth *(toile cirée)*, fpotted with black, became very white, and fpotted with brown fpots.

Fluid vegetable alkali, being the foluion of blue potafh expofed to the fimple contact of the oxygenated muriatic acid, acquired the property of bleaching like the true water of Javelle, but, inftead of the clear yellow colour it at firft poffeffed, it became white and limpid, The bottom of the faucer made ufe of was lined with an infinite variety of very white cryftals, in thin brilliant plates, of a dry appearance, like talc or mica; having the appearance of fo many fections of the cryftals of fulphate of potafh, through the whole length of the prifms, terminating in their pyramids. Thefe cryftals might be one line and an half in length, one in breadth, and near a quarter of a line thick. This experiment, and another mentioned in the following chapter, feemed to prove that the violet colour of the lees, diftinguifhed by the name of javelle, is, as Berthollet has obferved, more particularly owing to manganefe, of which the colouring matter is carried off with the gas that efcapes.

The folution of mineral alkali, extracted from the foda of Alicant, and of an amber colour, being

being expofed in the fame manner as that of the vegetable alkali, acquired the fame property of bleaching, without, however, entirely lofing its own colour, or prefenting any cryftalization.

Pure water, expofed in the fame manner, obtained the fame property of bleaching, preferving its natural colour, without exhibiting any obfervable peculiarity.

Thefe three different fluids, by becoming thus impregnated by the muriatic acid gas, feem to prove that it is not abfolutely neceffary to agitate the water of the veffels to concentrate the gas. An experiment with the intermediate tubulated veffels of the old apparatus, in which I have obtained pure folutions of this gas, coloured yellow or greenifh, and marked from ten to twelve degrees of concentration, appear likewife to fhew that agitation of the water is not, in ftrictnefs, abfolutely neceffary.

Thread, which had been fubjected to the lees, and was merely moift, or flightly humid, with the lixivial folution, being fimply expofed to the vapour or oxygenated acid gas, acquired a ruddy white colour fimilar to that of the third immerfion, and without any kind of alteration.

Coarfe thread, macerated feveral days in a weak folution of fulphate of potafh, became three fourths bleached, and with much uniformity or evenneſs of colour.

Flax macerated in the ſame manner likewiſe obtained a very fine white.

Flax, macerated for ſeveral days in the ſolution of potaſh, one degree below zero, and expoſed, like the objects above mentioned, to the oxygenated muriatic acid gas, became of th moſt beautiful white.

All theſe different articles were ſubject to no alteration. It is true, that, being apprehenſive leſt the gas, with which they were impregnated, ſhould alter their texture in conſequence of its concentration when they ſhould become dry, I was careful to waſh them out in a large quantity of water.

May we not infer from theſe various trials, which were all made during the winter of 1790, that it is highly probable that threads and piece-goods might be advantageouſly bleached by ſimple expoſure to the vapour of the oxygenated muriatic acid. For this purpoſe it appears to me, that the various articles, ſlightly moiſtened with water or with lees, would require to be hung up in a very cloſe chamber, like that which is uſed for expoſing goods to the vapour of ſulphur, into which room the extremity of the diſtilling veſſels muſt be introduced, to to convey the gas in proportion as it ſhould be diſengaged. An experiment of this nature would require peculiar management, and its ſucceſs would be of the greateſt importance to the manufacturer.

CHAP.

CHAP. XXIII.

On the Possibility of applying the Residues to Profit.

THE residues to which the attention of the operator may be directed, in order to derive advantage, are: 1. Those of the retorts, bottles, or other distilling vessels: 2. Those of the immersions, or bleaching liquors: 3. Those of the alkaline lees, or soap: and, 4. Those of the baths of sulphuric acid.

The residues of the retorts, bottles, or other distilling vessels, are reducible to the following: 1. Manganese not discoloured, and the common muriatic acid coloured by manganese, if the muriatic acid has been used instead of the muriate of soda: 2. Sulphate of soda, and a small portion of muriate of soda not decomposed, if this last has been made use of: 3. Sulphate of potash, if lees have been used to extinguish the suffocating odour of the residue of the solution, which is always more or less impregnated with oxygenated muriatic acid.

Though

Though I have reduced the proportions of manganese to one-sixth less than directed by Berthollet, it is not, neverthelefs, difcoloured after the operation, or rather, it is only difcoloured very flightly, and in few places. In this ftate it ftill preferves fufficient virtue, that is to fay, enough of vital air to be mixed with about one-third of new manganefe of the fame quality. This property, or ftrength, cannot, however, be afcribed to manganefe in lumps, or interfperfed with quartz, though well cleared of foreign matter. The manganefe cryftallifed in needles, fuch as is fold by Lapelletier, has alone afforded me this very perceptible difference *. Every other manganefe, on the contrary, that is to fay, the fpecimens in lumps, afford a much lefs quantity of gas, and render the bleaching liquor lefs ftrong. This laft kind of manganefe is alfo harder and more troublefome to pulverife.

The manganefe taken out of the retort, after the firft diftillation, preferves almoft the whole of its metallic brilliancy, and foils the hands as before, and may be ufed to purify glafs. It is true, that in this ftate it feems to have increafed in bulk. Manganefe entirely decompofed is known by the whitifh or pale

* This manganefe is brought from Hambourg, in the duchy of Deuxponts.

purple

purple colour, which the ſtrong impreſſion of the fire has given it.

The ſolution of the reſidue of the diſtilling veſſels diluted with water, the evening after the diſtillation, is found on the following day, if the veſſels have been cloſed, to be of a fine red, inclining to violet or purple, accordingly as the ſolution has been more or leſs diluted; but this colour does not fail to diſappear by expoſure to the open air, or by the heat employed to evaporate it. In either caſe, the violet colour of the ſolution is changed for a ſhade inclining to apple-green.

It ſeldom happens that the water which holds the reſidue of the retort or bottle in ſolution, is not ſufficiently concentrated to afford, after remaining for a day or two in the receiving veſſels, cryſtals of the ſulphate of ſoda; but theſe cryſtals, which are of different ſizes, are covered with manganeſe, from which it is neceſſary to clear them. This is eaſily done, by putting a ſmall quantity of theſe reſidues into a veſſel, and pouring a little clean water upon them, which, after briſk agitation, muſt be immediately poured off, before the manganeſe ſubſides, into a proper veſſel intended to receive this laſt ſubſtance. This manœuvre is to be repeated

peated four or five times as quickly as poſſible, in order that leſs of the ſalt may be diſſolved.

This trouble of waſhing may be avoided, if the violet-coloured water, which covers the reſidue of the diſtilling veſſels, be carefully decanted off into wooden or leaden receptacles appropriated to this purpoſe. The cryſtals, which are ſoon afterwards formed in this water, are neat and clear as they ought to be. But it is neceſſary, after having decanted this violet-coloured fluid, that common water ſhould be poured into the retorts or bottles, for the purpoſe of facilitating the extraction of what remains. This, together with the water, muſt be reſerved by itſelf. If it be propoſed to ſeparate the mangaueſe, for the purpoſe of uſing it again, as I have before mentioned, the following method muſt be recurred to. The reſidue muſt be waſhed repeatedly with a large quantity of water till it gives no perceptible ſaline or acid indication. The reſidue muſt then be dried, and afterwards mixed with new mangaueſe, in the proportions before directed. If the waters of the waſhing be ſufficiently impregnated to render it proper to mix them with the violet water, in order to increaſe the product of cryſtals, whether by inſenſible evaporation, or by the aſſiſtance of heat, this muſt be done, taking

taking care only, that in the latter procefs leaden veffels muft be ufed, becaufe copper, iron, and moft other metals would be fpeedily corroded and deftroyed.

The fulphates of foda and of potafh, which are obtained from the wafhings of the refidue of the diftilling veffels, have not hitherto been applied to any ufe in the arts. It is poffible, as I have before remarked, to employ them for difcolouring certain ribbands, and effacing writing from paper or parchment, as well as for fcouring copper and iron for braziers, &c. Both thefe falts are likewife ufed in medicine when purified; but it may be doubted whether the apothecaries would purchafe them, becaufe the very fmall quantity they confume is afforded very cheap from the falt-works of Lorraine and other places.

It would, therefore, be much more interefting to decompofe thefe falts, and obtain the alkalis, in a difengaged ftate, which might, in that cafe, be ufed, to make the lees in the fubfequent operations. Berthollet, in the firft volume of the Annales de Chimie, informs us, that feveral perfons have communicated different recipes to him for effecting this purpofe; it were much to be wifhed that the authors would benefit the public by a more liberal communication.

cation. In the mean time I ſhall here remark, that it is very poſſible to decompoſe theſe neutral ſalts by means of liberal ſulphur and the muriate of ſoda, by the ſulphuric acid, and more eſpecially by certain metallic oxydes, particularly that of lead. I have ſucceſsfully tried this laſt method in 1784, which was indicated by Scheele. The alkali which is obtained by theſe different proceſſes is of the pureſt kind, and I have had reaſon to be aſſured, that, with proper treatment, it affords glaſs equal in beauty to flint or cryſtal glaſs *.

The ſecond reſidue, which may be applied to uſe, is that of the exhauſted liquor of immerſion. After the vital air, or oxygen, has been exhauſted, the odorant liquor contains nothing but muriatic acid and water; the liquor without ſmell likewiſe contains muriate of potaſh. This ſalt, as well as the neutral ſalts, with

* The Committee of Public Safety publiſhed, in the ſecond republican year, the various proceſſes for decompoſing muriate of ſoda, which it had received from the different authors or inventors. Eſtabliſhments may, therefore, be made for ſupplying the national commerce with the alkaline ſalt of ſoda, the uſe of which is indiſpenſible in different works, ſuch as thoſe of glaſs, ſoap, dyeing, bleaching, &c. for the ſupply of which ſeveral millions are annually expended among foreigners. *Note of the Author.*

A copy of this report may be ſeen in the Annales de Chimie.—*T.*

a fixed

a fixed alkaline baſe, is of ſome uſe in medicine, but it is not worth while to extract any thing but the ſulphate of ſoda. This may be decompoſed for the ſake of the alkali, if the reſult ſhould be attended with ſufficient profit. I ſhall ſimply remark in this place, that theſe exhauſted bleaching liquors may be effectually uſed in making ſal ammoniac. The different trials I have made on this ſubject, by combining them with the volatile alkali of putrified urine or rotten vegetables, have conſtantly tended to confirm my opinion. Laſtly, if it ſhould be found advantageous to reduce the pure bleaching liquor without potaſh to the merchantable ſtrength, it may be uſed for the ſubſequent diſtillations, in the ſame manner as other muriatic acid, inſtead of the muriate of ſoda and ſulphuric acid; unleſs, indeed, it ſhould be thought more advantageous to uſe it for making white lead or verdigris, both which combinations I have made and uſed in painting with ſucceſs. The verdigris might alſo be uſed in dyeing.

I have alſo occaſionally uſed theſe waters of immerſion of the muriatic acid without ſmell, to make the ſecond lees for piece-goods and threads. This fluid becomes as highly charged as if the lees had been pure. The exhauſted bleach-

bleaching liquor may likewiſe be uſefully employed in the firſt maceration of goods; for which purpoſe, when it is not highly charged with colouring matter, it is no leſs valuable than the new liquor from the pneumatic veſſels.

There is another property of the exhauſted bleaching liquor, which is, perhaps, of conſiderable importance, namely, that of accelerating the vegetation of plants; from repeated trials I can affirm that it poſſeſſes peculiar properties in this reſpect. I have at different times uſed it, inſtead of common water, on cauliflowers, chervil, peas, cabbages, leeks, &c.: and theſe various plants have not only grown more quickly than others of the ſame kind planted in the ſame bed, and watered with river water, but have likewiſe acquired double the ſize.

Beſides the property of accelerating vegetation, theſe waters have likewiſe the property to drive away, at the inſtant of pouring on the ground, the ſpiders, ants, worms, ſnails, and other reptiles of this kind, which are noxious to plants and ſeeds. A gardener, near the laboratory where I made the muriatic acid for bleaching, was ſo fully convinced of the advantage of theſe waters, from his own experience, that he requeſted, as a favour, that I would reſerve them for his uſe; and was continually

S ſpeak-

ſpeaking in praiſe of the good effects it produced on the plants in his garden.

But in proportion as the ſmall quantity of oxygenated muriatic acid, diffuſed through the exhauſted water, is of advantage to vegetation, ſo much more noxious it is to plants when in the form of gas or vapour. Plants expoſed to this elaſtic fluid inſtantly fade and periſh. I have frequently ſeen this effect on the plant monk's-hood, and even on vines, the leaves of which ſoon became yellow, and the ſtems, after having languiſhed for a certain time, partly died.

With regard to the third reſidue, of which the waters of lixiviation form a part, I think I have ſaid all that is neceſſary in the chapter upon lixiviums. I ſhall here only add, that if there were an opportunity of diſpoſing of them to advantage to a ſaltpetre-work, it would probably be more advantageous than to reduce them by evaporation. There is, however, reaſon to think, that the old lees might be reſtored to a certain point by boiling them a long time with lime; this earth, having the property of deſtroying the vegetable parts which cover and weaken the alkalis, might, perhaps, produce the ſame effect as reducing the ſolution to the ſolid conſiſtence.

The following is likewiſe an economical method

thod of conſtantly applying the ſame lees to uſe, which I have often employed with the greateſt ſucceſs. It conſiſts ſimply in throwing the aſhes, from which they have been extracted, into the fires uſed for domeſtic purpoſes in the houſe, ſuffering them to dry, and afterwards wetting them with the exhauſted lees from time to time, which are to be reſerved for this purpoſe. The flame of the wood, burned in the chimney (for theſe obſervations are only applicable to a wood fire), and the heat of the hearth, ſoon burn the impurities, which coloured the alkali, and the aſhes ſpeedily become proper for lixiviation as before. This operation, which demands very little care, may be of great uſe, even in the domeſtic concerns of a houſe where alkaline lees are uſed.

The ſoap-waters likewiſe are not to be neglected. It would be poſſible to decompoſe them, either by means of the waters which have ſerved for the baths of ſulphuric acid, or with thoſe of the exhauſted bleaching liquor; but the beſt uſe would be for the manufacture of ſaltpetre, for which purpoſe the alkali muſt be extracted by calcination. In the laſt caſe the proceſs is nearly the ſame as with the lees; that is to ſay, when the ſoap-water is reduced to the conſiſtence of extract, and nearly dry, the oil muſt be burned off in an open fire, which will leave the alkali ſoluble in water, and ready for uſe, in the diſtillations and lixiviations, in the ſame manner as new potaſh. I have practiſed this method, and muſt here remark, that new ſoap-water riſes in froth above the veſſel when it boils, whereas that which has been uſed does not exhibit the ſame property.

With regard to the baths of ſulphuric acid, which compoſe the fourth reſidue, when they are too much diluted with water, from the immerſion of wet articles, the ſhorteſt method is to add more acid, or elſe to concentrate the fluid in the ſame manner as I have obſerved with regard to the ſulphate of ſoda, and other ſalts. For this purpoſe it may be concentrated to ſuch a degree as to be uſed again inſtead of common ſulphuric acid, or it may be uſed for making alum or ſulphate of ammoniac, by combination with the alkali of urine or putrefying vegetables.

CHAP. XXIV.

The Method of bleaching Hemp and Flax in the unmanufactured State, as well as Thread and Piece-goods, by the Assistance of Water only.

I HAVE long remarked, that the rags or pieces of unbleached cloth, which have been set to ferment in order to make blotting-paper, became white to a certain point, in consequence of being washed or soaked, either in heaps or under the mallets, for the purpose of destroying their texture. The washing, in these circumstances, becomes more easy on account of the fermentation, which opens the threads of the cloth, and the mechanical process of the cylinder, or mallet, which renders the colouring parts more easily detached, and in a certain degree dissolved. I attempted to imitate this fermentation, and solution of the colouring part of the thread, by washing in a large quantity of water. I made my experiment in preference upon flax. I first macerated it in pure river water, in a vessel, where I suffered it to remain till the ſurface of the fluid was covered with numerous bubbles. In this ſtage I turned it, and ſaw, with pleaſure, that its grey colour was changed to a light yellow. I then changed the water, firſt waſhing out the flax, and left it till other bubbles appeared, when I waſhed it again. At the ſecond waſhing, I obſerved ſeveral parts which were whiter than the reſt, and at the ſame time obſerved a conſiderable quantity of ſmall portions of grey and yellowiſh impurities, which detached themſelves from the filaments of the flax. I then waſhed it with rubbing, and was not a little ſurpriſed to obſerve the quantity of impurity increaſe, and the flax become whiter in proportion. Encouraged by the ſucceſs of this waſhing, I then plunged the ſame flax into warm water, to haſten the ſolution of the other colouring parts, which had immediately fixed themſelves on the flax, as ſoon as it had dried, after taking out of the veſſel. I then preſſed it in the water, which diſengaged an additional quantity of colouring parts, and the flax appeared much more beautiful. I did not carry this experiment further, becauſe the flax appeared clear and white, to as great a degree as I ſuppoſed it would arrive at by this method, for no more impurities were detached. Though it appeared to be white, when in the ſtate of diviſion,

ſion, yet in the maſs it ſtill preſerved a ſlight ſhade of yellow, which with a ſimple bath of oxygenated muriatic acid totally diſappeared, without the uſe of lees, or any other particular preparation.

This experiment perfectly agrees with an obſervation which may be daily made upon pieces of cloth which are ſubjected to the fulling-ſtock. Some of theſe pieces have holes in them; and in order that theſe damaged parts may not be enlarged by the proceſs of fulling, it is uſual to ſecure them by ſewing on a piece of brown linen cloth. I have remarked, not without aſtoniſhment, that theſe pieces of unbleached linen, after having remained in the water for two or three days, with the cloths to which they were fixed, and which were thus expoſed in order to clear them, either from the ſolution of ſoap, the urine, or the fullers' earth, became as white as if they had been paſſed through the lees, and expoſed alternately in the field for ſeveral months, or the uſual time employed in bleaching.

This reſult likewiſe agrees with the method in uſe in India, where, according to the relation of travellers, the natives bleach their fine cottons, which we receive from them, in no other way than by wetting and evaporation by the ſun, and expoſure to the dew, without the uſe of lees, or any other preparation.

All theſe experiments prove, therefore, that it would be very poſſible to bleach with water alone, if not piece-goods, at leaſt flax, in as expeditious a manner as can be deſired. This has, to a certain extent, been put in practice by a certain induſtrious individual in the town of Amiens, named Baſle. Without any knowledge of this man, or his method, but from the ſimple recital of his diſcovery, that he had bleached hemp in the ſtalk by water alone, I was tempted to make the trial. In conſequence I ſet to macerate in water, during for about a fortnight, a certain quantity of hemp ſtalks, which had been gathered about five or ſix months, and afterwards dried in a barn, without undergoing the proceſs of rotting. At the end of fifteen days the hemp had recovered its original verdure, that is to ſay, the appearance it had when firſt gathered. I rubbed them much under water, which diſperſed the green matter which appeared on the bark, and diſcovered the fibrous part, which had a pretty good appearance. I ſeparated this, and left it to ſteep for ſeveral ſucceſſive days in freſh water, after which I gave it another rubbing, and immerſed it for a ſecond time. It then appeared of a very beautiful white,

white, nearly the ſame as thread acquires by the old method of bleaching in the field, or the new proceſs with the oxygenated muriatic acid. This flax retained only a very ſlight tinge of a pale ruddy colour.

Theſe various experiments evince how important it would be to bring the ſteeping of hemp and flax to perfection, particularly of the latter article, which in the department of La Somme, whence it is watered only on the graſs; but the deſire of gain, which attends to the weight only, and not the quality, will ſcarcely permit the old method to be laid aſide. On the other hand, the bleacher, who is accuſtomed to uſe lime in ſolution, and even in ſubſtance—an ingredient which is, in ſome reſpects, rendered neceſſary by his intereſt, and the black tenacious colour of flax thus watered—might alſo, perhaps, be unwilling to abandon this practice. For the cheap price, which the uſe of this method enables him to offer, namely, 3 ſols an ell, without regard to the breadth, may ſecure employ, which would, perhaps, leave him, if he were to uſe another method, ſomewhat more coſtly, though at the ſame time in every reſpect beneficial for the merchandize and the proprietor.

CHAP.

CHAP. XXV.

The Method of Bleaching written or printed Papers and Rags, whether unbleached, dyed, or coloured.

THE following proceſſes are extracted from different memoirs addreſſed to the Committee of Commerce of the French National Convention; alſo to the Commiſſion of Subſiſtences and Proviſions, on the 24th Frimaire, the 15th Pluvioſe, and the 9th, 14th, and 21ſt of Germinal, in the ſecond year of the French republic.

Bleaching of old printed Papers, to be worked up again.

1. Boil your printed paper for an inſtant in ſolution of ſoda rendered cauſtic by potaſh. The ſoda of varech is good.

2. Steep them in ſoap-water, and then waſh them, after which the material may be decompoſed, or reduced to apulp, by the machinery of the paper-mill. The waſhing with ſoap may be omitted without any great inconvenience.

Bleaching

Bleaching of old written Papers, to be worked up again.

Steep your paper in a cold ſolution of ſulphuric acid in water, after which waſh them before they are taken to the mill. If the acidulated water be heated, it will be ſo much the more effectual.

Bleaching of printed Papers without deſtroying the Texture of the Leaves.

1. Steep the leaves in a cauſtic ſolution of ſoda, either hot or cold. 2. And in a ſolution of ſoap. 3. Arrange the ſheets alternately between cloths, in the ſame manner as the paper-makers diſpoſe thin ſheets of paper when delivered from the form. 4. Subject the leaves to the preſs, and they will become whiter, unleſs they were originally loaded with ſize and printers' ink. If the leaves ſhould not be entirely white by this firſt operation, repeat the proceſs a ſecond, and, if neceſſary, a third time. The bleached leaves, when dried and preſſed, may be uſed again for the ſame purpoſes as before.

Bleaching of old written Papers without destroying the Texture of the Leaves.

1. Steep the paper in water acidulated with fulphuric acid, either hot or cold. 2. And in the folution of oxygenated muriatic acid. Thefe papers, when preffed and dyed, will be fit for ufe as before.

The Method of bleaching Rags of the natural brown Colour for the Manufactory of white Paper.

1. Let the rags be opened or feparated from each other, after previous foaking or maceration for a longer or a fhorter time, according to their texture and quantity. 2. Give a lixiviation in cauftic, vegetable; or muriatic alkali. 3. Pafs them through the oxygenated muriatic acid, more or lefs concentrated with alkali. 4. Let the mafs be then worked for a fufficient time in the apparatus of the paper-mill, and it may be advantageoufly fubftituted inftead of that which is afforded by white rags.

The white colour will be ftill better, if, after the maceration, the rags be opened and fubjected, as ufual, to the action of the mill; after which the pafte itfelf muft be fubjected to one lixiviation, one immerfion, and a bath of fulphuric acid. The mafs being then well wafhed

and

and preſſed out, may be thrown into a trough to be manufactured.

Method of bleaching Rags, of all Colours whatever, in order to make white Paper.

1. Let the rags be opened, as before. 2. Steep them in the oxygenated muriatic acid. 3. If, as it commonly happens, the colour is diſcharged by this firſt immerſion, let theſe bleached and decompoſed rags be immerſed in water acidulated with ſulphuric acid. 4. Complete the diſorganization by the mallets or cylinders of the mill, after having previouſly well waſhed them.

If the colour ſhould not be ſufficiently diſcharged by the firſt immerſion in the oxygenated muriatic acid, which is very ſeldom the caſe, give them another alkaline lixiviation, and after that a ſecond immerſion in the oxygenated muriatic acid; after which ſteep them in water acidulated with ſulphuric acid, either hot or cold, the latter of which is the moſt active and effectual; and, laſtly, let them be ſubjected to the action of the mallets or cylinders.

Red and blue colours are moſt tenacious. With regard to black, it will be ſufficient if they be ſteeped after opening their texture, 1. In a diluted ſolution of ſulphuric acid; and, 2. In a ſolution of the oxygenated muriatic acid.

If

If the operator could know that theſe rags had been dyed in the raw ſtate, a ſtill more brilliant white might be obtained by following the ſecond method deſcribed in the preceding article. But it very ſeldom happens that coloured rags have not been bleached before they were dyed. The manipulations may be performed with ſufficient ſpeed to bleach at leaſt three thouſand pounds weight in the courſe of the day, without appropriating any extraordinary edifice or workſhop to this purpoſe.

Theſe new methods, if adopted in the preſent circumſtances (of France), will greatly contribute to prevent the want or dearneſs of paper or rags. The quantities of the reſpective materials cannot be preciſely directed on account of the difference of the veſſels, the papers, and the colours; but practice and attention have ſoon regulated theſe matters. In the applications of theſe principles, the republic of France have obtained two valuable advantages. On the one hand, a greater quantity of linen has been ſaved for the uſe of the hoſpitals; and, on the other hand, it has been more eaſy to reſerve the rags of ſuperior quality for paper-money and the purpoſes of trade, which require peculiar ſtrength to undergo the circulation.

CHAP.

CHAP. XXVI.

THE valuable product of vegetable alkali, which may be expected from the incineration of the marc or residue of grapes *, in order to shew the considerable magnitude of the advantage which this hitherto neglected material may afford, which in some vineyards is used as fuel, and in others as manure, I shall simply remark, that, after direct experiments, I find that five hundred pounds of the residue of this marc, dried after distillation (a purpose to which they are in some provinces applied to obtain brandy), and afterwards burned, have constantly afforded me one hundred pounds, or thereabouts, of ashes, which produced ten pounds of fixed vegetable alkali or potash, reduced to the consistence of saline of a blackish brown colour.

It is easy to perceive what an immense provision of this article one might annually obtain in countries where the marc is used only as manure.

* The following account is extracted from a memoir on the same subject, which was addressed on the 18th Thermidor to the Commission for Provisions and Ammunition to the armies of France.

manure. Nothing more would be required than to collect the ashes of this marc, burned after it is taken from the press, or subsequent to the time of attaining the small vines, or, lastly, after distillation in those places where they are used to make brandy.

The residue of grapes is not easily burned, because the seeds become disengaged. It may, nevertheless, be burned with a certain degree of speed, by means of a grate, having intervals of one inch, or half an inch wide, raised above an ash-hole or hearth to the height of 12 or 15 inches. The husks intended to be burned are to be disposed all round, because the previous drying is of advantage to hasten the combustion. The seeds which fall through must be thrown up, from time to time, with a shovel, until they are entirely red hot, in which state they are to be taken out and thrown in a heap in some convenient place near the furnace, where the combustion may be completed by turning them over from time to time, and exposing them to a current of air. The larger this heap is made, the more speedy and perfect will be the combustion, and the more abundant will be the alkaline produce. The heat remains a long time. I have seen instances in which the mass remained very red after having burned for a month

month with frequent ſtirring and expoſure to the air. It is proper to remark, that whatever care is taken to burn the ſeeds, there will, nevertheleſs, remain near one-tenth part not entirely conſumed, which may be eaſily ſeparated from the reſt with a ſieve. Theſe may afterwards be burned with the huſks, or ſeparately, as may be moſt convenient. The rain does not perceptibly injure the aſhes of theſe heaps, if they be covered with other huſks, either dry or ſlightly moiſt. If the marc be burned in this laſt ſtate, it produces a cinder, which is diſpoſed to agglutinate together in a ſtony form.

When the fire is once kindled upon the grate, it is afterwards kept up without intermiſſion, by charging it with dry huſks in proportion as the combuſtion proceeds. Theſe may be put on to the depth of ſix inches. When the fire is well lighted, and urged by a ſtrong wind, the moiſt huſks and ſtalks burn almoſt as ſpeedily as thoſe which are dry.

It is of eſſential conſequence, that the grates ſhould be fixed in a ſpacious airy place, where there is no danger from fire. The reſidue of grapes emits during its combuſtion a white and very thick ſmoke, which would incommode the neighbourhood.

Two grates of iron wire, about three quarters of an inch in the meshes, or holes, twelve feet in length, and four and an half or five feet wide, properly attended by one man, will burn in a day, by a fresh wind, five thousand pounds of this dry residue, which afford nearly one thousand pounds of ashes, and from the lixiviation of these ashes a product of one hundred or one hundred and ten pounds of well dried salin or vegetable alkali is attained.

The residue of grapes may also be made up in the form of peat, and dried in the open air, or under a shed. These are moulded in the same manner as the residue of the tanners. After drying for three days, they are sufficiently firm to be burned on a grate of bars one inch square, and an inch asunder, which form a kind of furnace either in the open air, or beneath a chimney. A furnace 8 feet long, 20 inches wide, and 18 inches deep, may consume four thousand of these dry pieces, each weighing about one pound and a half, and measuring in diameter five inches, and from an inch and a half to two inches thick. A woman, or a boy of fifteen years old, can make fifteen hundred of these cakes in a day with ease. This was the process I used in Messidor, in the year two, in order to afford

afford a ſpeedy ſupply of vegetable alkali to a manufactory of ſaltpetre, which I directed that year in the department of La Côte d'Or.

It may alſo be remarked, that advantage may very eaſily be derived from the heat of the fire, neceſſarily made of this proceſs, whether for the purpoſe of lixiviation of the aſhes, or drying the alkali.

The moſt eſſential fact, however, which requires to be known, is, that a very ſpeedy and abundant product of vegetable alkali, for the manufactory of ſaltpetre, or for other arts and manufactories, may be obtained by the ſimple means here pointed out.

CHAP. XXVII.

*Fabrication of crude Alkali (cendres Gravelées), with the Lees of Wine**.

UNDER the mantle of a kitchen, or a bakehouſe chimney, from one wall to the other, at the diſtance of 18 or 20 inches from the back, according to the opening of the flue, let a grate be fixed of bars of one inch ſquare, at the diſtance of about one inch and a half aſunder. This grate is to be raiſed at leaſt 18 inches above the hearth, and defended in front with a wall one brick thick, having perforations nearly reſembling thoſe of a pigeon-houſe. Or, inſtead of the wall, a grate may be ſubſtituted ſimilar to that of the bottom. This wall, or grate, may be 24 inches high. The interior ſpace of this kind of furnace, is then to be filled with the lees of wine preſſed dry or green. The latter are to

* The following is extracted from a memoir preſented on the 22d Vendemiaire, in the year two, to the Committees of Commerce and Proviſions, and to the Commiſſion of Agriculture and of Arts.

be

be preferred, becaufe the alkali they afford is much finer. The whole is to be fet on fire by means of ftraw or fmall wood, previoufly difpofed under the lees. The fire foon penetrates the whole mafs, and in lefs than a quarter of an hour the flame fpeedily reaches the upper ftrata, which muft be regularly fupplied with new lees, in proportion as the mafs finks down by the combuftion. That portion which falls through the grate, and may appear from its brown or blackifh fracture to be not entirely confumed, muft be returned again to the fire. The grate muft be cleared from time to time with a hook, in order to facilitate the combuftion.

Inftead of burning the lees in a furnace of this kind, another furnace may be ufed with equal advantage, entirely of brick, in the form of a hollow tower, at the bottom of which fome faggots of wood are to be placed, which are to be lighted after having filled part of the capacity of the furnace with frefh lees, mixed with dry, or with frefh lees only; for thofe which are too dry muft be fteeped a day beforehand, in order that they may be perceptibly moift. Thofe lees are then fucceffively thrown in at the top of the furnace, in which manner the procefs is to be continued until the whole of the lees are confumed. After continuing this procefs for feveral days, the furnace is ſuffered to cool, and the aſhes taken out through a door at the bottom.

It is proper again to remark, that theſe furnaces ought to be conſtructed in a ſpacious place, on account of the dangers of fire, and the inconvenience of the ſmoke, which is very conſiderable.

The lees, thus burned, emit a light very long flame tinged with different colours. It may be advantageouſly uſed in furnaces inſtead of wood, either alone, or mixed with that combuſtible.

One barrel or piece and a half of *cendres gravelées*, weighing about 260 Paris pounds, is the product per day by one man's work of the combuſtion of 6 or 7 barrels or pieces of wine lees, well dried in a furnace conſtructed under a chimney 20 or 24 inches deep, 18 inches wide, and 8 feet in length. This was filled four times during the day, taking care each time in the firſt place to ſtir the lees at top and in front with an iron fork, and to clear the grate beneath with a hook, which is abſolutely neceſſary to be done, in order to prevent the maſs from coagulating, and to favour the action of the air. With proper attention 50 or 60 barrels may be burned in ſix days, without being obliged to employ the night; the ſame quantity may be be burned in three days and two nights in the round furnace above defcribed, if conftructed of a diameter of 5 feet.

The good *cendres gravelées*, or that which is afforded by the red lees, contains at leaft 70 or 80 pounds of vegetable alkali in the hundred, when it has been carefully burned. That which is made with white lees, though well pounded and diffolved in hot water, does not afford more than 45 or 50 pounds of alkali. In general, when the cinders have acquired a green or blue colour in the fire, and are light and fonorous, the quality is good; but it feldom happens that the whole product is of the fame colour. That which is too much burned and refembles the fcoria of iron, muft be rejected, not only becaufe it is very difficult to pound, but likewife infoluble and earthy. The faltpetre-makers, dyers, bleachers, potters, leather-ftainers, glafs-makers, and others who ufe this faline fubftance, are particular in their choice of that which is light, fpongy, and of a greenifh or bluifh colour, which fhews no fign of vitrification in its fracture, the undoubted fign of too ftrong a fire.

CHAP. XXVIII.

The Method of Bleaching Wool.

PIECE-goods formed of thread and wool, or wool and cotton, being at present very much in fashion (in France), it may probably be an interesting object to manufacturers in this branch, to see an account of the method of bleaching wool in this place by the ordinary process, though well known. I shall in the first place speak of carded wool, which is used for broadcloths, and wool proper for combing, which is used for the manufacture of stuffs *.

Wool for carding for the manufacture of broadcloths, &c.—This kind of wool, as it is usually found in the market, has already been subjected in the hands of the grower or dealer to a cleansing, which has deprived it of 50 or 60 per cent. But the wools forwarded to the manufactories are still loaded with a portion of the unctuous or

* The following is, in great part, extracted from different memoirs of Roland Laplatiere, and Allard, formerly inspectors of manufactures.

greasy

greaſy matter, which it is neceſſary to clear them of. Experience has ſhewn, that this ſmall quantity of natural greaſe is neceſſary to preſerve them from worms, during the carriage, as well as the time of keeping, before they are applied to uſe.

The manufacturer's method of cleanſing his wool, is uſually performed as follows:—To a given quantity of water poured into a boiler, for example, five-and-twenty pails, an addition is made of five of old putrified urine, which is boiled for a ſhort time. The mixture is then proper to diſſolve the greaſe of the wool. Three or four pails of this liquid are then poured into a veſſel at the heat which will admit the workman to hold his hand in it. About 20 pounds of wool are then thrown in, which, after ſteeping a very ſhort time, are continually ſtirred with a ſtick for a quarter of an hour, after which it is taken out and drained for a few moments in a baſket over the veſſel, into which the drainage liquor returns. The wool is then carried to the river and repeatedly waſhed in large open baſkets, by ſtirring it about with long poles or rakes, till the water comes off very clear. In the mean time another workman puts a like quantity of wool into the veſſel, and the ſame operation is repeated, &c.

Several

Several essential matters are required to be observed during this operation. 1. The bath in the boiler, as well as in the other vessel, must be refreshed * from time to time, when its force is found to be diminished, for it loses much of its strength, in consequence of the various changes of the wool, as well as from evaporation. 2. If he should think the addition of urine unnecessary, by way of restoring the bath, he must raise its temperature, in order to give it the requisite power for the washing. 3. But the heat must be carefully prevented from increasing beyond the proper degree, as determined by experience, for it is found that too much heat hardens the unctuous matter, instead of dissolving it, and that too great a quantity of urine changes the wool, and renders it harsh. 4. When the bath has become too foul for the washing, it must be entirely changed: it may easily be imagined, that this work requires a very intelligent workman; but practice renders the business very easy.

The success of this operation is ascertained from the appearance of the wool, which becomes

* It is more advisable to refresh than to renew the bath, because the greasy impurity of the wool, which is disengaged by washing, becomes a leaven which disengages the grease from the other wool, plunged in the bath.

white,

white, soft, elastic, and open, dilating or swelling when touched, instead of being hard, greasy, and close, as it was at first. The qualities it acquires in the bath do, therefore, sufficiently shew the necessity and utility of this second cleansing, by which it loses 10 or 12 per cent more. This last loss, added to the former, gives a total of about 60 or 70 per cent; that is to say, 100 pounds of raw wool produce scarcely more than 30 or 40 pounds in a very clear state, fit for the manufacturer.

Wool for combing for the manufacture of Stuffs. —— This wool, in the market, is broken or sorted by the clothier, and sent before or after the dying (if this be intended) to the combers in parcels of about six pounds and a half each. The quantity is first washed in a vessel filled with hot water, taken out of a small boiler in which two or two and a half pounds of green or black soap has been dissolved, for the said quantity of wool, which accordingly as it is thought to be more or less foul, is well pressed and afterwards wrung on the hook, and then dried in the sun, or in the open air. Before it is combed, it is again subjected to a second bath of the same kind. These two clearings are sufficient to deprive it of all the natural grease which remains, and

and of ſuch impurities as might be an impediment to the combing proceſs.

It muſt be remarked, that theſe ſix pounds and a half of wool are waſhed in ſucceſſive ſmall portions at a time. The water of the waſhing-tub is renewed as the work goes on, in order to detach the greaſe and other impurities from the wool; there are two hooks fixed within the veſſel, one at each end, one of which can be turned round by a handle. The workman, after having well waſhed and preſſed with his hands the ſeveral parts of the wool, wraps them round the two hooks, and by wringing it out, he expreſſes the dirty water, which carries with it all the greaſe detached by virtue of this ſtrong preſſure. After this ſecond waſhing, the wool is dried carefully to prevent its being accidentally ſoiled.

In this ſtate it is, that the wool is combed. It muſt be rather moiſt for this operation, in order to facilitate the prolongation of its filaments, of which, when the wool is well cleanſed, the comber ought always to form lengths of three or four feet each. It is, therefore, eſſentially neceſſary, that this operation ſhould be well managed, not only for the good effect it produces in the opening, but, likewiſe, becauſe the

colour

co our and clearneſs of the ſtuff depends much more upon this firſt operation, than it is generally imagined.

In many manufactories, after the wools are combed, and according to the kind of ſtuff intended to be made, it is uſual, in order to diſpoſe them to ſpin well, to give them a third waſhing in the ſame veſſel with hot water and ſoap. The wool is afterwards carefully dried, and in this ſtate delivered to the ſpinner, if it be intended for the chain or weft; but that which is intended for weft is returned to the comber, and after coming out of his hands it is waſhed a fourth time as before. But this fourth and laſt waſhing is not given except to wools of the firſt quality, manufactured of a white colour, or intended to receive any clear and brilliant dyes.

Wool, which is well cleared of the greaſe, ought to have its filaments ſlender, long, even, and not connected with each other, beſides which it ought to be tenacious, white, and diſengaged from every foreign ſubſtance. The wool from Holland is remarkable for this laſt quality. That of England, is harſher and much fouler. The German wool is ſtill harſher, but equal to this in length. It approaches the French wool, which

which is the worſt kind of any, with regard to its length and fitneſs for combing.

The loſs ſuſtained by cleanſing, is ſomewhat leſs than one-fourth in the Dutch wools, and about a fourth in thoſe of England. The German wools, and thoſe of France, undergo a ſtill more conſiderable loſs, on account of their inferior quality. Some of the latter loſe more than one-third.

Sulphuring. Wool, ſtuffs, ſtockings, and other articles of the ſame nature which are ſoiled by dreſſing or uſe, are expoſed to the vapours of ſulphur. By this proceſs theſe goods receive a clearer white than that which is natural to the wool after the uſual waſhing and cleanſing.

This operation is uſually commenced by waſhing or fulling the piece. For this purpoſe it is requiſite that the fulling rammers ſhould be made lighter than uſual. When the convenience of a ſtream is not to be had for moving them, it will be ſufficient if a frame of 15 or 20 inches wide be made with two beams three or four inches thick, ſupported by croſs-pieces, and terminating below in a croſs-piece ſomewhat longer, ſtronger, and vertically ſuſpended to a plank or poles placed between the timbers of the roof, and forming a ſpring. A wooden trough is placed underneath, in which a workman may move the ſpringing peſtle up and down with his hand with great facility, and by inclining the trough the ſame effect of turning the ſtuff may be produced as in the common fulling apparatus.

Inſtead of a machine of this kind, the manufacturer may uſe the mallet, or which is ſtill better, the goods may be worked with the feet, in a place properly diſpoſed for this effect, as has been recommended for piece-goods and ſtockings.

When the piece is well cleanſed and rinſed in a ſtream, it is dried and ſinged, or ſent to the dye-houſe; if, on the contrary, it is intended for a clear white, it muſt be ſinged before the ſcouring *. For the fine white, a ſecond ſlight waſhing is given in a ſolution of ſoap, in which the ſtuff is left for a certain time, then waſhed well, rinſed in running water, and left to drain for an hour on the horſe, after which it is expoſed to the vapours of ſulphur for five or ſix hours, or longer, as far as 24 hours, according to the bulk of the piece.

After this operation, it is again waſhed, and its

* The method of ſinging muſlins, is equally applicable to woollen goods which require this treatment.

its colour heightened with fine whiting and blue, which are diffused in clear water; it is then sulphured a second time, washed in a slight solution of soap, dried, passed through the stretching machine, callendered, or pressed, according to its nature.

The following is the method of treating a piece of cloth of 40 or 42 ells, with the whiting and blue. Seven or eight pounds of fine whiting *(blanc d'Espagne)* are pounded and mixed up with water in a pail. This mixture, except the coarse particles at the bottom, is poured into a small trough of clear water. The bath being well mixed, the piece is passed rapidly through it upon the reel for a quarter of an hour, after which it is raised out of the bath upon the reel, and a pail of water is added, in which an ounce and a half of the finest indigo, or Prussian blue, has been diffused by the usual method of pounding, sifting, and wrapping it in a bag. The bath being again well stirred, the piece is immediately returned through the fluid again by means of the reel. After this treatment, it is laid on a packing-cloth, and carried to the workshop, where the nap is laid by the fullers' thistle, during which the surface is wetted with the fluid of the bath, and when the piece is dry,

it

it is beaten with twigs to clear it of the white powder it received in the foregoing procefs.

It is proper to obferve, that bad fmells, and even the offenfive breath of individuals, will fometimes produce a change in the bath of blue and white, in which woollen goods are fteeped; or, at leaft, this is what very refpectable manufacturers affirm to be the cafe. When this happens, the operator is obliged to plunge his piece in a bath of hot water, to wafh out the white and the blue, which have fixed themfelves irregularly in a kind of vegetation, after which the operation muft be repeated. With regard to woollen fhirts, flannels, and other articles intended to be worn next the fkin, neither fulphur nor foap are in any refpect fuitable to them. It is fufficient if thefe be well fcoured in bran and water, and afterwards well wafhed in clear water. The colour is of no particular confequence, as the main object is to render it as abforbent as poffible, to which quality the foap itfelf is a great impediment.

The preference is often given to leave ftockings on the leg with wafhing or fulphuring them.

The place in which the operation of fulphuring is performed, is merely a very clofe chamber, in which the goods are fufpended on poles of white deal, ſo as to hang down in folds, which neither toucheach other, nor the floor or wall. It is ſtill more particularly neceſſary, that they ſhould not touch any iron, which becomes oxyded by the muriatic and the volatile ſulphureous acid afforded by the ſulphur which burns in a veſſel on the floor, and would certainly ſpot it. Inſtead of paſſing theſe pieces over the poles, it might, perhaps, be more adviſeable to faſten them beneath the ſame poles, by means of hooks paſſing either through the liſts themſelves, or through loops of twine attached to the liſts.

It is neceſſary to be aware, that a cloth which has undergone the operation of ſulphuring, ſhould not be immediately laid upon wood before it is purged of the ſulphureous acid, which would diſſolve the reſinous or gummy parts, and ſpot the goods.

The ſulphuring not only communicates a diſagreeable ſmell to the cloth, but likewiſe gives it a harſh feel. A bath of ſoap which is given after this operation reſtores its ſoftneſs, and that in a degree which is more effectual the longer the cloth is worked in it.

CHAP. XXIX.

The Bleaching of Silk.

THE ſame reaſons which have led me to inſert the proceſs of bleaching wool in the foregoing chapter, with the account of the goods which are wholly or in part made of that material, induced me likewiſe to inſert the proceſſes for bleaching ſilk.

There are two methods of performing this, either by ungumming it, or leaving the gum in its texture. I ſhall treat of both, beginning with that in which the ſilk is ungummed and boiled *.

This proceſs is managed as follows: Diſſolve, in a ſufficient quantity of water, in a boiler over the fire, 30 pounds of white ſoap of Marſeilles for every 100 pounds of ſilk. After the ſolution has boiled, lower its heat by an addition of cold water. Extinguiſh or ſlacken the fire, but take

* Here, as in the foregoing chapter, I recur to the Memoirs of Roland Laplatiere.

care, neverthelefs, to keep the bath at a confiderable heat. Steep therein the filks, hung on rods, in which ftate leave them till their whitenefs and flexibility fhews that the gum is diffolved and feparated. Spread out the filk on the rods, and turn them, in order that the parts out of the bath may be fteeped in their turn, and when each hank is perfectly ungummed, wring them on the pin to exprefs the foap; fhake them, and put them in bags of coarfe cloth, containing 20 or 30 pounds each.

Make a new bath in the fame proportion, and in the fame manner, as the former. Throw the bags therein, and boil them for an hour and a half, ftirring them from time to time in the boiler. The ungumming and boiling of filk deprives it of 25 per cent of its weight.

If the filk be intended to be dyed, the ungumming and boiling are performed in the fame bath, which is boiled for 3 or 4 hours, making ufe of a quantity of foap proportioned to the finenefs of the colour, or rather the white ground which it requires; 25 or 30 pounds are fufficient for common colours, and as much as 50 for thofe with faffranum, and poppy red, cherry colour, &c.

But when it is intended that the filk fhould be white, and, confequently, to bleach it, the bags

bags are carried to the river, when they are taken out of the boiler, and the ſilk being taken out, is extended upon cords floating on the water and well waſhed.

A new bath, containing a pound and a half of ſoap to 30 pails of water (of about three Engliſh gallons each), in which a ſmall quantity of litmus, with a portion of powder-blue or indigo, is diffuſed, according to the nature of the ſhade intended to be given. The boiler is filled, the bath heated, but never to boiling, and the ſilk is paſſed through it over the rods, until it has uniformly acquired the requiſite ſhade. It is then wrung dry, and hung out, or elſe carried to the ſulphuring room.

All the ſilks made uſe of in the white, in any manufacture whatever, require to be ſulphured in order to bleach them more perfectly. One pound and a half or two pounds of ſulphur are ſufficient for one hundred pounds of ſilk. At the expiration of 24 hours, the room is ventilated, and muſt not be entered until the vapour of the ſulphur is diſſipated. The air which enters in ſummer is ſufficient to complete the drying of the ſilk, but in winter this is performed by a chafing-diſh or ſtove put into the room.

If the white or ſulphured ſilk ſhould not prove blue enough, a new ſhade is given with clear

 water;

water; the hardeſt water is beſt, after which it is ſulphured a ſecond time.

With regard to ſilks intended for gauzes and blonds (one of the principal qualities of which is derived from the natural rigidity of the ſilk), they ought not to be either ungummed or boiled. The whiteſt natural ſilks are choſen in preference, which are ſteeped and opened in a bath of clear hot water, or ſoap and water. In the firſt caſe, they are wrung, and afterwards ſulphured. The fine ſilks of Nankin, which are of a beautiful white, have no need of this operation.

The following is the method publiſhed by Rigaud in 1778, for bleaching ſilks without ungumming them *.

The ſilk, intended to be bleached, is put into a glaſs veſſel containing a mixture of ſpirit of wine and muriatic acid, in the proportion of a pound of the former to half an ounce of the latter,

* This method differs a little from that publiſhed in 1793, by Baumé. See this laſt, Journal de Phyſique of the ſame year may be conſulted, and for that of Rigaud, the Gazette du Commerce of the 7 Novembre, 1778. *Note of the Author.*

This method requires many precautions, and would be much too expenſive if the materials were not afterwards recovered. An abridgement of Baumé's paper, which contains a detail of theſe objects, may be ſeen in Nicholſon's Philoſophical Journal, I. 11. 32.—N.

latter, and, in quantity, ſufficient to float the ſilk. The veſſel is then cloſed with wet parchment, and expoſed for 12 hours to the ſun, or otherwiſe it may be left 24 hours in the ſhade, at a temperature between 15 and 20 degrees of Reaumur. The ſilk is then taken out and preſſed, and again macerated for the ſame time, and under the ſame circumſtances, in freſh acidulated ſpirit of wine, in another ſimilar veſſel cloſed as before. The ſilk is then taken out, preſſed, and waſhed for four or five minutes in pure ſpirit of wine. In the next place, it is kept for 24 hours in the ſun, or 36 in the ſhade, in a third veſſel, containing pure ſpirit of wine, which is to be renewed at intervals, after which the ſilk is to be taken out, preſſed, and waſhed two or three times in clear water, which is to be changed at each waſhing. Laſtly, the ſilk is expoſed to dry upon a frame ſo contrived as to ſtretch it with conſiderable force, and prevent its curling up as it dries.

EXPLANATION OF THE PLATES.

The ſame Letters denote the ſame Things in the correſpondent Figures of Plans, Elevations, Sections, and Profiles.

PLATE THE FIRST.

FIGURES 1, 2. 10, 11. The plan, ſection, profile, and detail of a diſtilling apparatus, entirely mounted and ready for ſervice. It ma ybe formed either with a ſingle or double apparatus, as is ſhewn in the figures. Each apparatus is compoſed of two ſeparate furnaces, which are, neverthelefs, ſupported on the ſame ſtructure, with two diſtilling-veſſels, one pneumatic veſſel, and a veſſel for immerſing the goods.

A. A ſtructure of light wood-work, which ſupports the furnaces or their maſonry in brick or plaiſter.

B. Platform of brick or tile, ſerves as a hearth to the aſh-holes, C, of the furnaces.

D. A bed of clay, on which the brick platform is ſupported.

E. The wood-work, or planks of the veſſels in which the bed of clay is placed.

F. A vacant ſpace, in which the mixtures of muriate of ſoda and mangaueſe, in the proper doſesf or changing each diſtilling veſſel, are kept

kept dry, for the purpofe of procuring the oxygenated muriatic acid by the muriate of foda, inftead of directly ufing the common muriatic acid at 25 degrees of concentration of the areometer of Moffy.

G. The door of each of thefe receptacles.

H. The drying pans. Thefe are a kind of troughs or capfules of plate-iron of a fquare form, in which the muriate of foda is put to dry, either before or after pounding and fifting.

I. The ventsor chimnies through which the fmoke or fumes of the coal efcapes, which is ufed in heating the diftilling veffels.

J. The chimney of the furnace, leading under the drying place.

K. A capfule or veffel of plate-iron, either fquare or cylindrical, for the purpofe of fupporting the diftilling veffels and the fand-bath in which they are placed. It is moft adviseable to form this veffel cylindrical, becaufe the flame, in that cafe, applies better to its external furface, and a lefs quantity of fand would be required to be heated.

L. The door of the furnace.

M. A ledge or ftep, fixed to the frame of the furnace, in order that the operator may raife himfelf upon it fufficiently high to pour the mixtures into the diftilling veffels, or for any other operation relative to the furnaces.

N. The

N. The diſtilling veſſel, or retort, having its neck O, and its adopter P, which laſt may be of glaſs, ſeparate from the retort, or elſe a part or prolongation of the ſame, ſuppoſing the glaſs-men to be ſufficiently ſkilful to give it this figure. In order to obviate the accidents of fractures, it may be made of ſtone-ware, porcelain, or, which is ſtill better, of lead, as is ſhewn in the figure,

Q. A welt or projection of lute which fixes the adopter to the retort. Inſtead of the retort, the operator may uſe, with ſtill greater advantage, balloons, or tubulated veſſels, ſuch as are deſcribed in plate 9, fig. 1 and 2. I prefer theſe laſt veſſels becauſe leſs expenſive, more common, more generally uſeful, and, in particular, more convenient. Q. 1, a pipe of glaſs, ſtone-ware porcelain, or lead, the latter of which is preferable; its extremity, Q 2, is fitted to the adopter; and its other extremity, Q 3, ſuffers the oxygenated muriatic acid to eſcape in the form of bubbles into the pneumatic veſſel.

R. The pneumatic veſſel, placed on its three-legged ſupport S.

T. The arbour of the agitator. U its fans, or arms. V. Handle for turning it. X. Diaphragms or falſe bottoms, beneath which the oxygenated muriatic acid gas is concentrated and abſorbed; they are ſupported on one ſide by the regular inclination

clination of the ftaves of the veffel, and on the other by the pegs of wood Y: thefe falfe bottoms divide the pneumatic veffel into a number of feparate receptacles.

Z. The pipe through which the gas paffes from one cavity to the other; its prolongation prevents the gas from immediately efcaping into the upper-chamber; the gas being by this means forced to remain for a time in the inferior chamber, where it is frequently agitated by the arms of the apparatus, becomes abforbed in the water to a certain degree. &c, a funnel of wood to facilitate the pouring of water into the pneumatic veffel, when its cover is fixed on, pinned faft, and the places fecured by paper pafted on.

a. Spigot, or cock, to draw off the acidulated water for trial of its ftrength, by the known re-agents, indigo or cochineal, as mentioned in chap. 14. This cock may be formed of glafs, or lead, or even copper; but this laft metal muft be covered with a coating or two of white lead paint, to prevent its being rufted, or oxyded by the vapour of the gas, and its confequent fpotting the various goods which may come into contact with it, or may be foiled by the falling of particles of verdigreafe with which it would become covered.

b. The cocks, for emptying the bleaching liquor

quor into the veſſels of immerſion: they ought to be of wood, cloſed either with a cork, or with a turned pin, ſecured with flax; they muſt likewiſe be firmly fixed in the pneumatic veſſel, and well defended with fat lute, within and without.

c. A tube of glaſs, of the ſize of barometer tubes; that is to ſay, 2 or 3 lines in diameter: it ſerves to ſhew the height of the liquor which remains in the veſſel, when a portion has been drawn off for particular immerſions; and it likewiſe indicates the greater or leſs action of the diſtillation, by the frequency with which the liquor oſcillates up and down without. This laſt indication is particularly uſeful toward the end of the operation, when the ſlowneſs and weakneſs with which the bubbles eſcape, produce ſcarcely the leaſt ſound in the veſſel, even though the ear be applied to its ſides.

This tube is fixed at the diſtance of about an inch from the bottom of the pneumatic veſſel; its place of junction is well ſecured with fat lute, within and without; its upper extremity is ſecured in its place, by a ſmall piece of wood, *d*, pinned to the pneumatic veſſel.

e. A ſhort tube, of about the length of two inches, below each inferior falſe bottom; it does not ſuffer any gas to eſcape from one cavity to the other, excepting that portion which cannot incorporate

corporate with the water, either becauſe it may already be nearly ſaturated, or becauſe too large a quantity may be collected in the upper part of each cavity, reſpectively, for want of the agitator being worked with ſufficient frequency.

f. Pipes of lead, or ſtone-ware; they may likewiſe be made of wood; one of them paſſes through all the falſe bottoms, the other paſſes only through the uppermoſt; they ſerve to introduce, upon the bottom of each cavity, either the quantity of potaſh neceſſary to fix the odour of the muriatic acid, or that which may be neceſſary to form the liquor, known under the denomination of the oxygenated muriate of potaſh.

Theſe two pipes are cloſed during the diſtillation, with a ſtopper of cork; their upper extremity, being formed ſomewhat in the ſhape of a funnel, facilitates their ſuſpenſion and ſupport; they muſt be well ſecured with fat lute, at the place of contact, with the partition through which they paſs.

g. The cover of the pneumatic veſſel. It muſt be kept in its place by good pins of oak formed with heads, and its joints all round muſt be cloſed with ſtrips of paper paſted on The joints of the ſeveral pieces or planks which compoſe it, though tongued together, are likewiſe covered with paper: by means of theſe precautions, no ſmell of the

the oxygenated acid exhales. In order likewiſe that this vapour ſhould not eſcape through the ſmall ſpace between the cover and the arbour of the agitator, this laſt is ſurrounded with a ſmall quantity of flax, or piece of rag, wetted, either with common water, or a ſolution of pot-aſh. When the diſtillation is begun, the opening of the aperture of the funnel *t* muſt be cloſed with a cork.

h. Supports of the adopter of the retort: they reſt on the cover of the pneumatic veſ-ſel, and receive the upper extremity of the tube, communicating with the internal part near the bottomof the veſſel; this extremity is diſpoſ-ed in the form of a funnel; 1. To receive the beak of the ad: pter; 2. To facilitate the applica-tion of the lute. The two parts which compoſe the ſupport are connected together, either with Iron-wires, or pack-thread, or elſe by means of hooks. *i* a hole to ſuffer the air to eſcape out of the pneumatic veſſel when the water is pour-ed in.

l. The veſſel for immerſions, mounted on its rollers J. *m* the reel to move the piece-goods in the liquor. *n* its handle. *o* the piece, or good , paſſing over the reel. *p* the dotted lines, repreſenting the cover. It may conſiſt either of a cloth, thrown over the veſſel after the liquor has been poured in, or it may be much

more

more ſuitably and conveniently formed, by means of two frames of light wood, with panes of glaſs, which open on the oppoſite faces, and cloſe towards the upper part of the triangle, which they form by their junction. In order that the odour of the liquor may not be inconvenient to the workman, at the time it is poured into the veſſels, in thoſe caſes, where it is not thought proper to extinguiſh it; either by a certain doſe of ſifted chalk, or a proper addition of ſolution of pot-aſh in water, one of theſe frames has a proper opening to ſuffer the diſcharging cocks to paſs through; and thoſe parts of this opening, which are not accurately cloſed, are afterwards ſecured by means of cloths tied round the cock itſelf. In order, moreover, to avoid all ſmell from the pipes, communicating from the bottom of the pneumatic veſſels to the veſſels of immerſion, theſe may be ſo diſpoſed, as to convey the bleaching liquor to the bottom of this laſt, and cauſe it to riſe gradually, inſtead of pouring it in with agitation.

Fig. 3.—Perpendicular view of the grate, or chafing-diſh, upon which the coal and charcoal *a*, which heats the capſules and retorts, is placed. They may be raiſed higher or lower by placing them on one or more bricks. It is likewiſe very eaſy to take them out of the furnace by means of the handle *B*, when it is required, from any cauſe

whatever,

whatever, that the diſtillation ſhould ſpeedily be checked."

Fig. 4.—The elevation in perſpective of this grate.

Fig. 5.—Plate of iron, which ſerves as a door to the furnaces.

Fig. 6.—The ſame plate, or door, ſeen in profile; *a*, a projection which ſerves to raiſe it, or place it; *b*, borders, by means of which it reſts on the joints, formed by the upper bricks of the furnace, which, at the time of its conſtruction, are left open for this purpoſe.

Fig. 7.—Supports which ſurround the upper extremity of the tube of lead, which receives the beak of the adopter of the retort, or the retort only, if the glaſsman have made it all of one piece.

Fig. 8.—Elevation, in perſpective, of the iron-trough, which contains the mixture of muriate of ſoda and manganeſe, which is ſet to dry, as before deſcribed, between the furnaces, under the drying place, during the time of diſtillation, in order that it may be ready for the ſubſequent proceſs.

Fig. 9.—Plan of the ſame trough.

PLATE THE SECOND.

Fig. 1.—A machine for grinding the glaſs ſtoppers of veſſels and bottles, which are required to be cloſed, as it were, hermetically.

A. A bottle with three necks.

B. The ſtopper fitted to cloſe the middle neck.

C. A kind of brace, one extremity of which has its centre of motion in the wooden ſlider E, and the other extremity F receives in its ſocket G, the ſtem H, of the preſs I, the two jaws of which, I, K, hold faſt the knob of the ſtopper, required to be ground in. The ſlider of wood E is not fixed, but muſt riſe or fall according to the wear of the ſtopper in the neck of the bottle, in which it is intended to be fitted.

Fig. 2 and 3.—Plan and profile of a veſſel, for immerſing thread ſtockings, or other pieces of frame-work knitting; if, inſtead of the round figure, this veſſel had been made ſquare, the goods might have been ſtowed in a more advantageous manner. Three baſkets of white wicker-work may be placed one upon the other, as is ſhewn in figure 2, each upper baſket reſting on the handles of that beneath. B, a bundle, or hank, of ſkains of thread ſpread out in the bath: there muſt not be more than three or four of theſe connected together in the ſame bundle, as, otherwiſe, they would be leſs conveniently expoſed to the action of the bleaching liquor, and leſs eaſily wrung or cleared out. In order that no dirt may fall into the veſſel, and to prevent any oppreſſive vapour from incommoding the workmen, the whole is covered with a piece of cloth, designed for this purpose; or, rather, with a light covering of wood, cut into two pieces, to facilitate the management.

Fig. 4 and 5.—Plan and elevation of a small portable boiler for the purpose of immediately boiling the thread in alkaline lees, or any other small articles, which either cannot with convenience, in point of time, or otherwise, be put into the large boiler with piece-goods, or other articles, whether on account of their fineness, their small quantity, the speed required, or their being the object of some particular experiments.

a. The boiler, placed on its tripod, *b*, under the mantle of the chimney; *c* its cover, which must never be neglected to be put on, not only because the heat is more speedily produced, but likewise for the purpose of defending goods from soot, which may fall down the chimney, and would produce spots not easily discharged, as has been mentioned in chap. 15. This boiler may be heated by means of wood, or turf, or pit-coal, if a proper grate may be made use of.

Fig. 6.—The method of suspending, by means of an arbor D, the basket, containing the articles taken out of the boiler, and draining over it. In order that no impurities may fall either into the

lees,

lees, or upon the goods, during this operation, it will be proper that a cloſe cloth, or frame of light wood, ſhould be ſupported in that part of the pipe of the chimney which is immediately over them.

e. The baſket, or plat form, of oſier, ſupporting the goods; this plat-form may likewiſe be made of iron, covered with linen rags: if a baſket be uſed, it muſt be perfectly cleared of its bark, for fear of ſpotting the goods.

The four cords *f*, which ſupport the baſket, are united in pairs, on each ſide, at the knot *g*, to the cord *h*, which winds on the arbor.

j. Supports, between which the arbor turns. *k*, a clump faſtened on the arbor, to prevent its recoiling from the ſupport *i*, in conſequence of the friction upon *m*. *n*, the handle of the arbor.

Fig. 7.—Rollers, for the purpoſe of folding piece-goods into lengths, after they have received the dreſſings.

a. Cords, one extremity of which is attached to the ring *b*, fixed to the cieling, and the other extremity bears the two gudgeons of the wooden roller *d*.

e. Part of the piece of cloth, to be folded in two.

f. Part of the cloth already folded. In this

 operation,

operation, which is very ſpeedy, the cloth is ſupported by holding one part, *f*, in one hand, and the other part, *e*, in the other, taking care to bring the edges together by raiſing this part of the cloth thus joined a little, the weight of the folded part, ſoon draws over that part to the other ſide of the roller, a new portion is ſucceſſively folded, and is thus ſubjected to the ſame manipulation.

Fig. 8, 9.—Plan and profile of a machine for folding cloths in equal folds, whether it be required that the folds ſhould be in the whole length, or that they ſhould be folded in two, as has already been obſerved.

a. Uprights of iron, placed oppoſite each other, in the holes *b*, in the braſs *c*, fixed on each ſide of the table *d*.

e. Rods of iron, or wood, placed in each fold of the cloth behind the two uprights.

f. Another rod, which raiſes from the heap of cloth, *g*, that part which is to be carried behind the upright, over the rods. In proportion as the folding advances, the lower rods are drawn out to be uſed in the progreſs of the work. By this means the operation may be performed with a dozen rods. The uſe of theſe rods, and the regularity which they afford in this method of folding, which is very expeditious, has cauſed it

to

to be named folding by the rod, in the ſame manner as that deſcribed in figure 7 is named folding by the roller.

Fig. 10.—End view of the manner in which the pieces are folded together after they are taken from the machine, fig. 9: the round fold, which is given to the piece, is ſecured by means of ſtrings, more or leſs fine, *a a*, according to the quality of the merchandize. Theſe cords, ſtrings, or twiſt of ſilk, or flax, paſs from the upper ſelvedge to the lower; they are faſtened together with a knot, which is, laſt of all, covered with a taſſel, *b*, of coloured thread, of ſilk, or linen, or thread, according to the beauty and fineneſs of the piece. Coarſe goods are likewiſe faſtened in front, as is ſhewn by the letter C.

Fig. 11.—Hanks of linen or cotton thread; the union of ſeveral ſkains, that is to ſay, five or ſix by a ſingle ſtring, *a*, forms what is called a hank: by the number of knots, *b*, made in one or other of the two ends of the ſtring, the bleacher diſtinguiſhes the merchant to whom the article belongs. For this purpoſe the diſtinctive ſigns are entered in a book, oppoſite the name of the proprietor. With regard to other articles, ſuch as piece-goods, ſtockings, &c. they may be diſtinguiſhed by one or more coarſe threads ſewed to them, upon which any number of knots at pleaſure may be made. This method of marking goods appears to be much preferable to the different marks with crayons, red oaker, &c. which always, more or leſs, ſpot and ſoil the goods.

Fig. 12.---Shews the method of rinſing the ſkain on the pin to clear it of water, lees, or any other fluid it may contain. As the part which is neareſt the pin is not ſo effectually wrung as that which lies near the middle, care muſt be taken in opening the ſkain, *c*, to replace it in ſuch a manner that the part which was upon the pin, *a*, ſhall be near the middle at the time of the ſecond wringing. After this ſecond wringing, the thread is, for the moſt part, dry enough, and may be ſtraitened out; which is done by paſſing it over the hand, and ſtrongly jirking or ſhaking it by drawing out; or the operation may be performed with the wringing-pin inſtead of the hand.

Fig. 13.—The method of diſpoſing the ſkains, ſo that they may not intermingle too much with each other, particularly thoſe of ſewing thread, which, inſtead of being wrung on the pin, are worked under the lower part of a kind of rammer or ſtirrer. The four faſtenings, *a*, prevent the ſkains from becoming intermingled during this kind of fulling proceſs. A tub or pail may be be uſed for this operation, according to the quantity of the article, intended to be cleared in this manner. A man or woman may work them with the inſtrument, fig. 14, without much difficulty: a certain degree of ſkill and intelligence is required to do the buſineſs in the moſt effectual manner.

a. Repreſents the handle of this inſtrument, and *b*, the lower part, which is made of beech cut into ſteps or notches, *c*, on each ſide, in order that it may take a ſlight hold of the goods, and afford a facility in turning them. Theſe indentations muſt be rounded at the edges and corners.

Fig. 15, 16.—Plan and elevation of a waſhing implement, with which ſtockings, thread, and other ſmall and fine articles, may be cleaned or rinſed in a tub or pail. *a*, exhibit the teeth or feet, between which the different goods diſpoſe themſelves, and are agitated againſt each other, for the purpoſe of clearing them of the different liquids, with which they are ſucceſſively penetrated during the courſe of the bleaching. *b*, is a double handle, by the aſſiſtance of which this ſmall inſtrument is moved.

Plate 3. Fig. 1.—*a.* Birds-eye view of the great boiler, in which piece-goods are heated in the alkaline ſolution or lees.

Fig. 2.—Section of the fame boiler through the line A, B.

a. Floor of the fhop.

b. The mafonry, in which, the copper C is fet: *d*, the wooden cover, formed of two or more parts. *e*, crofs pieces which pafs under the cramps *f*, fixed to the boiler itfelf, which keep down the covering and prevent it from rifing by the force of the ftream, which laft is, by thefe means, compelled to re-act on the pieces of cloth, or other articles placed in the boiler.

g. A cavity or gutter furrounding the copper, and ufed for evaporating without any other heat than that which it receives from the boiler itfelf, the old lees, which are referved after they have been applied to every ufe, which can be made of them, in order to recover the alkali.

h. A fmall boiler, heated by the heat which efcapes from the great boiler, before it paffes into the chimney. In this fmall boiler is kept a quantity of lees, ready prepared, of the proper ftrength. It here receives an increafe of temperature, which prevents its fenfibly retarding the boiling of the great boiler, when it is neceffary to convey a certain quantity into this laft veffel.

i. A cock, by which the pure folution of alkali is drawn off, and may be conducted to the

the great boiler, by a channel of tin or wood, &c. &c.

j. The fire-place under the boiler. Inſtead of the iron-bars, or a grate, which, on account of this diſtance between them, ſuffer too much air to paſs, for the conſumption of wood, and by theſe means waſte much of that fuel, I have preferred an arch of brick-work, with openings of a ſufficient ſize. This arch, while it ſaves the wood, likewiſe preſerves the heat, on account of the ſpace between the vents, upon which the burning fuel remains for a longer time. It might perhaps be poſſible to diminiſh the expence ſtill further, by having no apertures at all through the arch. Theſe apertures are alſo liable to be worn or broken, and require to be defended at their edges with iron.

k. The aſh-hole.

l. The chimney, proceeding from the fire-place, immediately beneath the ſmall boiler.

n. Stairs aſcending from the floor, to the brick-work of the floor.

o. Another ſet of ſtairs, leading to the platform *p*.

q. A regiſter for regulating the heat of both boilers.

r. Flue of the chimney.

Fig.

Fig. 3.—Section of figure 1, along the line *c d.* *a*, the floor. *b*, the masonry in which the copper is set. *c*, mouth of the fire-place of the great boiler. *d*, lower part of the chimney. *e*, ash-hole. *f*, register, to regulate the heat of the small boiler. *g*, the stairs from the platform of the masonry to the floor of the work-shop. *h*, stairs leading to the upper part of the masonry.

Fig. 4.—A crane, moveable on its axis, provided with tooth and pinion-work, by means of which the charge of the boiler, consisting of piece-goods, or other articles, may be raised. *a*, the shaft. *b*, the pivot. *c*, the arm. *d*, bracket, or support. *e*, a ratchet wheel, worked by a pinion with a double lever: round the barrel of this wheel is wound the chain, at the other end of which are three or four hooks, in which the chains *b* are held. These last are covered with cloth or cord, to prevent the effect of rust. The intermediate part between the two concentric circles, which form the vessel, or suspended apparatus, is likewise defended by small cords, in order that the various articles disposed therein may not escape; this stage, charged with the different articles which have undergone lixiviation, is, by means of the crane,

crane, conveyed over other veſſels, where it is lowered down upon croſs pieces, in order that the drainage may be completed.

Fig. 5, 6, 7.—Plan, ſection, and elevation, of an oven for calcining crude alkali, in order to convert it into potaſh. *a*, ſtairs, which lead to the back part of the oven, where there are placed two boilers of caſt-iron, *b b*, in which the alkali is dried, after having been concentrated to a certain degree in the cavity ſurrounding the great boiler. See Fig. 1, 2. Theſe two boilers may be appropriated alternately to dry the alkali entirely, whence it is to be conveyed into the calcining oven; and afterwards to complete the reduction of the concentrated alkalies to the conſiſtence of extract; and may likewiſe be diſpoſed in ſuch a manner that the flame which they receive from the fire-place of the oven, before it paſſes into the lower part of the chimney, may heat a third boiler of caſt-iron, of copper, or even of lead, which may be uſed to evaporate the old lees, or ſolutions of ſoap: for theſe laſt require the ſame management in order to obtain their alkali.

This concentrated alkaline ſolution from the upper boiler, may be ſuffered to fall, drop by drop, into the lower, in order that the evaporation,

tion, or complete drying, may not be impeded by too much water being ſuddenly poured in.

c. Paſſage from the fire-place to the ſpace beneath the caſt-iron boilers. It conveys a ſtream of flame, which is more than ſufficient, and may be governed by means of regiſters. As this paſſage is conſtructed on a ſlope, which, for that reaſon, is not eaſy to be made in the maſonry, a contrivance was uſed during the building of the roof of the fire-place, to fix in the proper place a roller of wood, upon which the bricks in part reſted which were intended to form this paſſage; it may eaſily be imagined that this wooden roller could not be taken out after the conſtruction was finiſhed; it was, therefore, intended that it ſhould be burnt out. To accelerate the combuſtion at that time, and during the heating of the oven, a hole of two or three inches in diameter was bored through it. This piece of wood may be of white deal, or any other material which is moſt readily conſumed. The heat conveyed by this paſſage, of which I have aſcertained the good effect by experiment, is very well regulated by means of a regiſter made at the bottom of the ſpace immediately beneath the boiler. This regiſter is entirely ſhut when the boilers are not intended to be uſed.

When

When the calcining oven is not uſed, but the boilers are wanted to dry alkaline ſolutions, theſe laſt may be ſeparately heated by a fire between both.

d. An aperture in the roof of the oven, through which the dried alkali is conveyed into the oven.

e. The calcining oven. In order that the alkali may be eaſily diſengaged from the edges or angles of the floor where the walls take their riſe, at which place it readily fixes itſelf by the aqueous fuſion, it is adviſeable that this part ſhould be defended by a plate of iron, four or five inches high, and about half an inch thick. By this means the ſalt is more eaſily ſeparated by the rake.

f. The ſtoke-hole for ſupplying the fuel. It is conſtructed in the ſame manner, and for the ſame uſes, as that of the boiler for lixiviation.

g. A ſlight piece of brick-work, between the fire-place and the hearth of the oven, which prevents the fuel and the ſaline matter from communicating or mixing with each other.

h. A ſtone or caſt-iron trough, into which the red hot calcined alkali is thrown when taken out of the oven. In this veſſel it is left to cool before it is packed up in caſks, in which laſt veſſels it muſt be preſſed as cloſely as poſſible, in

order

order that it may be less liable to attract moisture from the air.

i. The mouth of the oven. It has two iron uprights, *j j,* forked at top, in which the ends of a racked bar *k* are placed. The handle of the rake, with which the alkali is stirred, is rested between the notches of this bar. The mouth of the oven is also provided with an iron plate, to be used either for closing it entirely, or more or less, as occasion may demand.

Instead of suffering the heat, which issues from the mouth of the furnace, to be lost, it may be very advantageously directed by a pipe rising under the boilers of cast-iron, or those which are placed above, for preparatory evaporation. None of these means ought to be neglected of employing the heat, which in almost every construction of this kind has been hitherto lost, for want of a proper degree of skill in the proprietors, or those who undertake to erect them.

Fig. 8.—Represents part of the chain, which is wound upon the drum of the ratchet-wheel of Fig. 4. It may be observed, that it is constructed on the same principles as the chains of pocket watches.

Fig. 9, 10, and 11.—Details relative to the racked

racked bar placed acroſs the mouth of the calcining oven.

a. The teeth, between which the handle of the rake is moved. *b*, an elbow, which prevents the rake from moving the bar ſide-ways when once duly placed. *c*, crampons, or iron fixed pieces, which receive and ſteady the uprights. *d*, the maſonry of the oven in which they are placed. *e*, the rake ſeen ſide-ways. *f*, its claw placed on the floor of the oven. *g*, its iron handle. *h*, the external part of its handle, which is made of wood, becauſe the iron would communicate the heat too readily to the hand of the workman. Fig. 11. the claw of the rake ſeen in face. Fig. 12. an hook, by means of which the rake is lifted up or changed for another, either on account of its being too hot to be touched, or in danger of bending. *a*, the hook. *b*, its handle. In order that the iron handle of the rake may ſlide more readily between the teeth of the croſs-bar, it is occaſionally rubbed with a piece of bacon nailed to a ſmall piece of wood.

PLATE THE FOURTH.

Fig. 1.—Elevation of a mill proper to cleanſe piece-goods and other articles, which are more or or leſs bulky. It is moved by a horſe; or its mechaniſm may be ſet in motion by wind, water, or other powers, by making ſuitable arrangements for that purpoſe.

A, the principal ſhaft. B, the bar, to which the horſe is attached. C, the wheel. D, lantern. E, the arbor, which gives motion to the ſtocks or peſtles E, by means of certain tripping pieces. See Fig. 1. and 2, Plate V. F, mortices, in which the tripping pieces move to raiſe the peſtles. It is adviſeable that the part which is acted upon by the tripping piece ſhould be defended, either by a plate or roller of copper. G, croſs-pieces, between which the peſtles riſe and fall. H, the box or receptacle, in which the goods are placed. The figure repreſents the internal part, in conſequence of a portion of the wood-work in front being removed. One of the ſpaces is larger than the other, for the purpoſe of ſubdividing the different kinds of work. The receptacle is commonly made of elm, and the rammers of beech. I, holes made at different heights, to draw off the water. J, a door, which may be taken down or put up at pleaſure by means of buttons. It muſt always be up during the time of work. K, a pipe, which ſupplies the work with water. Out of this proceed a number of ſhort

ſhort pipes anſwerable to the reſpective chambers. L, a ſtop, to prevent the peſtle from deſcending too low, and injuring itſelf. M, a lever fixed to a croſs-piece, N, behind the four uprights, O. By means of the pin P, and the ſtring Q, the fulling rammer may be raiſed ſo high, that the tripping pieces cannot reach it. While it remains thus fixed, the goods may be examined or taken out, as occaſion may require.

Fig. 2 and 3.—A ſucking-pump, which furniſhes the water to the work. Fig. 2, ſhews the ſame in profile, and Fig. 3, the front view.

A is the extremity of the arbor which works the fulling apparatus. B, prolongation of the axis or pivot of this arbor. It turns on the braſs bed C; and its extreme part D is bended into a handle, which gives motion to the pump, rods E F. The latter is attached to the ſtem of the piſton which moves in the body of the pump. H. I, the pipe, which ſupplies the reſervoir J with water. K, the pipe which conveys the water to the fulling works.

PLATE THE FIFTH.

Plan and elevation of the ſame machine for waſhing and cleanſing piece-goods. A, the turning arbor, with its tripping pieces B. C, the mortices, in which the tripping pieces enter to raiſe the hammers. D, holes, through which the rammers traverſe. E, bolts, which hold together the lateral partitions, and connect them with the uprights F. G, the fulling rammer, reſting on its ſtop H. I, the lever, to throw the fulling rammers in or out of work. The dotted lines J denote the poſition of the levers when the work is ſtopped. The cord K being faſtened to the hook L, keeps the lever in this poſition; but when M is fixed to the ſame hook, it keeps the lever clear of the working bar. N, the inſide of the receptacle for the goods. O, holes for drawing off the water. P, the door. Q, the pipe which brings water to the work. R, a table, on which the goods are laid before or after they have been put into the engine. T, a board to defend the workman from being wetted.

Fig. 2.—A, the principal arbor, with its tripping pieces B. This figure ſhews in what manner they are diſpoſed in order to produce their alternate effects on the rammers.

Fig. 3.—Plan of the wheel fixed on the upright ſhaft, which ſerves to ſhew the manner of its conſtruction.

Fig. 4.—The lantern which moves the arbor. A, repre-

A, repreſents the arbor itſelf, upon which the lantern it ſolidly fixed. The bolts B connect the two drum-heads.

Fig. 5, ſhews the method of diſpoſing the goods in the trough of Fig. 1, when it is more particularly intended to work them acroſs their length. Fig. 6. The method of diſpoſing them when, on the contrary, it is intended to work them in the direction of their length.

PLATE THE SIXTH.

Fig. 1, 2.—Plan and profile of the machine for ſquaring and rolling out the pieces after they have received the dreſſings. *a*, the ſtage on which the goods are placed. *b*, the piece folded back and forwards. *c*, the ſtretcher, through the mortice of which the cloth paſſes. At one of its extremities there is a ratchet-wheel, *d*, by means of which the neceſſary tenſion is given, that the cloth may undergo a ſlight degree of friction againſt its rounded edges during its courſe. *e*, croſs-pieces, at ſuch a diſtance from each other, that the cloth, by paſſing alternately over one and under the other, may be gently rubbed againſt their blunted edges. *f*, another ſtretcher, through the mortice of which the piece likewiſe paſſes before it arrives at the wooden

 cylinder.

cylinder. This is likewiſe provided with a ratchet-wheel, *g*, for the purpoſe of ſtretching the cloth more or leſs. As the tenſion is conſiderable in this part, there is a lever, *h*, fixed on for the purpoſe of facilitating the turning. *i*, a cylinder or roller of wood, upon which the cloth is rolled, and left for a longer or ſhorter time, as may be neceſſary for it to keep the figure and dimenſions which it has received. The workman, who ſtands before this roller, takes care for this purpoſe to arrange and draw out the ſelvedges in ſuch a manner that they may apply at every turn upon the ſame parts of the cloth which are already rolled, and preſerve the ſame width throughout. *j*, a moveable piece, which may be thrown forward, and ſerves to keep the roller in its place endways, and when drawn back, leaves it at liberty to be taken out for the purpoſe of unrolling the cloth. A groove is made lengthways in the roller, for the purpoſe of fixing the end of the cloth therein, which is firſt wrapped round a wooden rod, and then lodged in the groove. *l*, braſs rollers, upon which the wooden cylinder turns. *m*, the ſquare, into which the ſquare end of the cylinder is lodged. *n*, wheel of the arbor, which carries the ſquare. *o*, the lantern or pinion, which gives motion to the wheel, and is itſelf

carried

carried round by the handle *p*, worked by one or two men, according to the force or velocity required to be exerted. *q*, a fly, armed with balls or plates of lead, which ſerves as a reſervoir of force, and greatly aſſiſts the workman. *s*, a trough of plate-iron, in which hot embers are put for the purpoſe of drying, or giving a proper degree of firmneſs to certain goods before they are rolled on the cylinder, upon which they preſerve the ſtate and appearance it is intended they ſhould receive.

Fig. 3 and 4.—Plan, ſection, and profile, of the earthen furnace, made *au rue Mazarin*, of which mention is made in Chapter II. *a*, the furnace. *b*, the aſh-hole. *c*, the door of the aſh-hole. *d*, the fire-place. *e*, door of the fire-place. *f*, grate of earthen-ware; inſtead of which, if preferred, a grate of iron may be ſubſtituted. *g*, the chimney. *h*, a protuberance for more eaſily removing the furnace. *i*, the pot. *l*, ſand-bath. *m*, tubulated bottle, containing the mixture for diſtillation. Inſtead of the bottle, a tubulated retort may be uſed, which, in that caſe, may be placed in a bath ſuited to its figure. *n*, the neck, to which the tube of lead is to be adapted, for the purpoſe of conveying the gas into the pneumatic veſſels. *o*, the aperture, into which the ſulphuric or muriatic acid is to be

 poured,

poured, accordingly as the diſtillation is performed with or without the muriate of ſoda. *p*, a ſtand or baſe of ſtone, upon which the furnace may be raiſed, either for the purpoſe of giving it a proper elevation, or to preſerve the floor from the danger of fire.

This kind of furnace is uſually compoſed of one ſingle piece; but for the facility of removing and fixing them, when conſtructed of a certain ſize, they ought to be formed of ſeveral pieces which may be eaſily fixed together by means of indentations made in them before they are baked.

PLATE THE SEVENTH.

Fig. 1, 2, and 3.—Bird's-eye view, elevation, and profile, of the machine for calendering piece-goods with or without heat. A, a double handle which gives motion to the pinion B. This machine, as well as the one juſt deſcribed, may eaſily be moved by connecting it with the fulling mill, in the ſame manner as the pump receives its motion, namely, by a branch or tumbler, which on the one hand is applied to the arbor of the mill, and on the other to the leading axis of the machine. It is neceſſary of courſe to arrange the workſhops accordingly. The pinion

pinion drives the toothed wheel C; on the axis of which is fixed the lantern or pinion P, which moves the great wheel E, to which is adapted the braſs cylinder F, and this in its turn communicates its motion to two cylinders of walnut-tree G.

H, the ſtage upon which the piece I is placed ready for calendering, having previouſly receiv-ed all the other dreſſings; it is folded, as the figure ſhews, in alternate folds, in order that it may be more eaſily delivered. It firſt paſſes between the croſs-pieces J, thence through the mortice K, of the ſtretcher L, which is provided with a wheel and click, M, to ſtretch the piece more or leſs and regulate its courſe. It after-wards paſſes back under the braſs cylinder N, over that of walnut-tree, and returns in front under the upper wooden cylinder, which it en-velopes as it paſſes over and falls behind O of the machine on the roller P, and againſt Q, where it is ranged in alternate folds on the ſtage R, whence it is taken to be regularly made up for ſale.

S. The preſſing ſcrew, by means of which the upper wooden cylinder may be urged more ſtrongly againſt that of braſs, accordingly as it is required that the face of the cloth ſhould be more or leſs acted upon.

T. Piece of caſt-iron, which ſlides in grooves made in the uprights U, and bears upon the pivot V of the upper cylinder, accordingly as the ſcrews preſs upon the croſs-piece X, to which this piece of caſt-iron is fixed.

Fig. 4.—The braſs cylinder ſeen at one end, where it is open to receive the bars of red-hot iron which heat it.

A. The cylinders of wood to which the braſs cylinder communicates its motion. Inſtead of wooden cylinders, others may be ſubſtituted of card-paper, compoſed of the quantity of leaves neceſſary to fill the ſpace which is determined between the plates of braſs adjuſted at the extremity of their axis. They are ſtrongly preſſed by theſe plates, which are retained in their poſition by powerful ſcrews. Cylinders of paper, properly turned, give to fine goods a glaze which they cannot acquire by the wooden cylinders. They have likewiſe the advantage of retaining their figure, which is not the caſe with wooden cylinders. Theſe laſt require to be occaſionally rectified in the lathe, and at laſt become too ſmall for uſe.

The lathe uſed for turning any of theſe cylinders ought to be conſtructed in ſuch a manner, as to render it a matter of certainty that the diameter

diameter ſhould continue equal from one end to the other.

B. The cylinder of braſs.

C. The neck on which it turns in the manner of a pivot.

D. The aperture through which the bars of red-hot iron are introduced with tongs. The aperture is then cloſed with a cover to keep in the heat.

U. The uprights between which the braſs cylinder moves againſt the plates V.

Fig. 5.—The form of the pieces of caſt-iron with which the cylinder is heated. Two are commonly put in, and they are uſually changed every hour, or oftener, according to the nature of the work.

PLATE THE EIGHTH.

Fig. 1 and 2.—Plan and ſection of a veſſel particularly deſigned for ſubmitting fine piece-goods, ſuch as muſlins, linens, &c. to the action of the oxygenated muriatic acid.

a. A frame armed on each ſide with ſmall leaden points or pins, *b*, the central parts of which are iron; they may be altogether of iron, painted with white lead, and well wrapped with ſtrips of linen or ſtring. Theſe points are of uſe ufe to fufpend the piece-goods in a zigzag form, as is fhewn by thefe figures, either immediately by the felvedge of the piece, or by loops of tape fewed to the pieces themfelves.

c. Rings of lead caft upon rings of iron, which they cover; otherwife rings of iron alone, painted with white lead, and wrapped with cloth or twine to defend them from the ruft. The cords *e*, which are faftened to thefe rings, pafs over hooks at *d*, fixed to the cord *f*, which, by means of the pulley *g*, is ufed to raife or lower the frame. When the latter is entirely plunged in the bleaching liquor, the cords are detached from the hook *g*, and the veffel is covered to prevent the introduction of any impurities, as well as to defend the workmen from any difagreeable odour.

An apparatus of this kind may be ufed, not only for immerfing of the goods in the oxygenated muriatic acid, but likewife for the lees, as well as the bath of fulphuric acid, &c.

Fig. 3 and 4.—Elevation and profile of the frame for ftretching the fkains of thread when taken out of the bleaching veffels. A, upright pieces, in which a number of holes, B, are pierced for the purpofe of fupporting C, the crofs-pieces, over which the fkains of thread, D, pafs fingly. Thefe crofs-pieces have their

angles

angles well rounded, and are ſet at a greater or leſs diſtance, according to the degree of tenſion to be produced, by means of the iron pins inſerted in the holes of the uprights.

The thread is firſt well wrung upon the pin, or with the hand, after which it is ſtretched upon the pannel, and left to dry. Care muſt be taken that thoſe ſkains which are put on at any one time ſhall all be of the ſame length, in order that the tenſion may be equal throughout.

If theſe threads, when taken out of the bleaching veſſels, ſhould be ſo entangled or mixed as to ſeem incapable of being cleared without breaking, they may eaſily be brought to their original ſtate by plunging them in water, and gently ſtriking them with the edge of the hand. This operation may, if thought fit, be repeated at each immerſion, in caſe it ſhould be ſuppoſed that the thread would be too difficult to clear after the entire bleaching.

The ſkains of thread, thus adjuſted and dried, are afterwards twiſted together in dozens or ſcores, or any other count, according to the cuſtom of the market; or they may be packed in ſmall parcels in coloured paper, according to their quality, and the orders of the vender.

Fig. 5, 6, 9, 10.—Plan, profile, and parts of the machine for ſcorching or ſingeing muſlins, and

and other piece-goods, which are particularly required to have an even face ſimilar to goods of the ſame kind imported from England.

A, poſts fixed in a cavity, B, in the pavement or floor of the workſhop. They are connected by the croſs-pieces, C, fixed at their upper extremities by the bolts or ſcrews D. Theſe poſts may be taken up at pleaſure, in order to leave more ſpace in the workſhop. E, axis of a roller placed between each poſt, the prolongation of which is bended into a handle, F, for regulating the tenſion of the piece by the ratchet-wheel G. A ſhort piece of cloth or canvas, H, is nailed to each of theſe rollers, and to theſe the extremities or ends of the piece to be ſinged are fixed with the needle or rod of iron, K. One of theſe rollers takes up that part of the piece which is ſinged, while the other gives off or unfolds a new length to be ſubjected to the ſame operation.

L, the bended iron, with which the ſingeing is performed by paſſing it over the ſurface, from one edge to the other, in a light, ſpeedy, and dextrous manner. The flat part M being firſt made very hot, is well wiped on a cloth or pad, in order that it may not ſoil or greaſe the piece, which laſt action would endanger its being burned. This iron is to be paſſed two or three times

times over the extended part of the cloth, in order that it may produce its effect more uniformly.

If the piece require to be finged on both fides, it muft be afterwards turned, by changing the rollers M for N. The brown fcorched colour which the piece acquires by this treatment fpeedily difappears in the bleaching.

Fig. 7 and 8.—A plate of caft-iron, viewed in face and edgewife, which is advantageoufly ufed to fmooth or lay the nap of certain piece-goods, fuch as velverets, fuftians, coverlets, &c. This plate is heated to the proper degree; and one or two men, according to the weight and dimenfions of the plate, and the nature of the ftuff, pafs it along, more or lefs fpeedily, over the piece intended to be thus ironed or fmoothed. The fame care muft be taken to wipe the plate when it comes out of the furnace as was defcribed for the fingeing. The handles, B, of the plate, are wrapped round with cloth, in order to defend the hands of the workman.

Fig. 14.—A brufh, with fhort hair, ufed to raife the fibres or nap of the piece-goods intended to be fubjected to either of the operations here defcribed.

Fig. 11, 12 and 13.—Plan, elevation, and fection of a prefs proper for drying or expreffing the water

water from piece goods, whether in the courſe of the ſeveral operations, or at the end of the bleaching proceſs. This preſs may likewiſe be employed to advantage, to give a neat appearance to folding goods, or for the cloſe packing of bales.

A. The platform or table of the preſs upon which the goods are placed. This table is hollowed out to the depth of one inch, and is inclined towards the gutter or beak, B, in order that the waters which flow from the goods may be received in a pail placed underneath for that purpoſe.

C. The preſſing ſcrews which riſe and fall between the uprights, D, in order that the whole ſurface of the table may be left clear.

PLATE THE NINTH.

Fig. 1 and 2.—Plan and ſection of the diſtilling apparatus mentioned at the commencement of this work.

A. The double furnace with its fixed grate. B a cover of plate-iron of a ſquare or round figure with flat edges; which may be raiſed or placed in contact with the furnace. C the ſand bath which ſurrounds the capſule; it ought to be very dry, fine and uniform. D, the cylindrical

bottle

bottle with two necks; one in the middle, E, to receive the ſtem, F, of the communication of the pneumatic veſſel, and the other for pouring in the ſulphuric acid, when muriate of ſoda is uſed, or the common muriatic acid, if that ſalt be not applied. Inſtead of the cylindrical bottle, a balloon, or globular veſſel with a neck on one ſide, may be uſed, as is ſhewn in figure 2. The choice of theſe veſſels will, in a great meaſure, depend on the ſkill of the workmen, at ſuch glaſs manufactories as may be in the neighbourhood of the bleaching works. In ſtrictneſs, the neck or tube on one ſide may be diſpenſed with, and it is here mentioned only becauſe it adds a convenience to the operation.

From theſe obſervations on balloons, it is eaſily ſeen, that thoſe who, in purſuance of the directions in the memoir upon bleaching, in the ſecond volume of the Annals of Chemiſtry, may have uſed the mattras, the neck of which ſooner or later muſt break at the termination of the chimney of the dome of the furnace there recommended, may afterwards uſe theſe globular veſſels with advantage.

J. The door or opening to the fire-place.

Inſtead of glaſs bodies, it might probably be of advantage to uſe leaden veſſels heated by a water-bath, or in any other manner which would

not

not endanger the fuſion of thoſe veſſels. This danger would be leſs to be feared when manganeſe and the muriatic acid were uſed alone; but veſſels of this kind would always have the diſadvantage of not allowing the operator to ſee what paſſed within them, as he may with glaſs, nor whether the veſſels were well cleaned. Theſe two inconveniences, which can hardly admit of any remedy, unleſs a glaſs could be fixed in the upper part, have hitherto prevented me from uſing them, though they might, probably, be attended with very conſiderable advantages.

K. The aſh-hole; this is cloſed in the ſame manner as the aperture of the furnace, by ſliding doors; or more ſimply as has been deſcribed in the explanation of the furnace exhibited in plate I.

G. A tube of lead formed all of one piece, without ſolder, and caſt at one ſingle pouring, or ſeveral ſucceſſive pourings like water pipes in copper moulds. It may alſo be caſt very ſhort at one ſingle pouring, provided it be made thick enough to be afterwards drawn out. Theſe ſucceſſive drawings reduce its thickneſs to a ſingle line, while its internal diameter is kept at ſeven or eight lines, which proportions are very convenient. This pipe is fixed in the pneumatic veſſel in ſuch a manner that it may be freely raiſed

raiſed and lowered in the groove L, made for this purpoſe in the piece which ſecures it on each ſide in its place, by means of the wooden pins M.

N. Another tube for a ſecond diſtilling apparatus, if it be thought fit to place one beſide that already formed, whether for the purpoſe of obtaining a high degree of concentration in the oxygenated acid, or great ſpeed in the compoſition of the gas, on account of the haſte required in the work.

O. Arms of the agitator, which may be ſloped off on both ſides. This form agitates the liquor, and facilitates the abſorption of the gas more effectually than any other.

P. A ſocket applied to the arbor Q of the agitator. If the pneumatic veſſel be conſtructed according to the repreſentation in the figure, and the explanation given in the chapter IV. the ſockets, R, here expreſſed become unneceſſary, as well as the inverted ſtages, whether they be made with hoops, as at S, or conſtructed of thin wood-like ſieves, as at T. They are placed in this drawing only to ſhew the manner of diſpoſing them in caſe they ſhould be preferred. The croſs-pieces, U, of the bottoms of theſe kind of veſſels, placed upon cleats fixed to the ſtaves of the veſſel, ſhew the manner in which they are to be fixed, with wooden pins, that they may not be ſubject to riſe, nor become looſe in any reſpect.

X. The cover of the pneumatic veſſel.

Y. A ſtool upon which the veſſel itſelf is ſupported.

Fig. 3.--Exhibits the manner in which the hoops of thin wood are joined for the purpoſe of forming the inverted veſſels, as well at their terminations, where one part overlaps the other, as upon the circumference attached to the bottom itſelf.

a. Wooden nails with heads. Oak is the beſt material. The extremities of the hoop of thin wood overlapping each other, and ſecured on each ſide in the joint of their ligature with wooden wedges.

b. The extremities of the circle of thin wood lying over each other, and confined on each ſide in the joint of their ligature with wooden wedges.

Fig. 4.—The manner in which the hoops are put on which hold the ſmall ſtaves of the ſecond conſtruction together. The circles *a* are kept together without binding, merely by a ſcarf or notch made in each end. Theſe hoops are ſtripped of their bark.

c. The

c. The ſtaves. Theſe, as well as the bottoms themſelves, may be made of yellow deal.

a. The ſcarfed ends of the hoops, which are turned inwards.

Fig. 5.—The method of diſpoſing the tranſverſal bars to which the bottoms of the inverted veſſels are fixed. *a* the croſs-bars. *b* the bottom of the inverted veſſel. *c* the ſtaves. *d* the ſcarfed hoops. *e* wooden pins which fix the bottom of the inverted veſſel to the croſs-bar. They are rivetted beneath, by ſplitting the lower point of the pin, and driving a wedge into the notch. *f* a ſmall block or cleat, fixed to the ſtaves of the caſk with wooden pins, *g*, driven ſlantways above the croſs-bar to keep it ſteady in its place.

Fig. 6.—The method of fixing the inverted veſſels, which have a border of thin wood like a ſieve. *a* the hoop or circle of thin wood pinned to the circumference of the bottom. *b* heads of the pins which fix the circular part. *c* boards making part of the bottom, but left of a greater length, in order that they may be fixed down to the ſupport *d*.

As theſe circles of thin wood are not likely to join exactly to the circumference of the bottom of the veſſel, they muſt be made good with putty, after previouſly ſtopping the larger vacuities with tow driven firmly in. Pitch may be uſed inſtead of putty, if it ſhould be found more convenient.

Fig. 7 and 8.—Section and plan of the veſſel for immerſing linen, hempen or cotton thread. It is placed on rollers or trucks, *a*, for the convenience of removal.

A. Bars with the corners rounded off, which are ſupported at each end in a notch or mortice, B, in the-croſs piece *c*.

d. Skains of thread, ſeparate or ranged in bundles of two or three only. The poſition of theſe muſt be changed once or twice during the immerſion, in order that the part which reſts on the bar may be ſubjected in its turn to the action of the oxygenated muriatic acid.

E. A pipe of lead or wood, through which the veſſel is filled from beneath. If the acid were to be poured upon the thread, it would bleach more ſpeedily at the upper part than elſewhere. Inſtead of one pipe there may be more, or the diameter may be ſufficiently large for the ſpeedy filling of this veſſel.

F. A funnel through which the bleaching liquor flows from the pneumatic veſſel. In ſtrictneſs it may be ſuppreſſed, and the apparatus ſo diſpoſed that the cocks from that veſſel may diſcharge themſelves immediately into the upper part of the pipe, which muſt be fixed withinſide the

the veſſel, in order that it may not impede the covering and uncovering. The cover ought to be compoſed of ſeveral portions or frames of glaſs, as was directed with regard to the veſſel for immerſing piece-goods; as by this means facility of management, and ſpeed in the proceſs, are inſured.

G. A ſmall door or ſliding piece, in which a plate of glaſs is ſet, through which it is eaſy to obſerve the progreſs of the work. This may be opened from time to time to examine the goods without raiſing the covers.

This method of diſpoſing the ſkains of thread in the oxygenated muriatic acid, is likewiſe applicable to their immerſion in the ſulphuric acid; when either of theſe liquors is exhauſted, it may eaſily be drawn off, either by means of ſyphons or of a cock placed near the bottom.

Fig. 9.—Section of a veſſel for immerſing piece-goods, to which a pair of hooks is affixed to wring out the pieces in parts at a time, particularly if of conſiderable magnitude.

A. uprights, which may be eaſily adapted to the two oppoſite ſides of the veſſel by means of the two clamps, B, which are fixed to the veſſel with a hinge on one ſide and a ſtaple and pin on the other or by ſtaple and pin on each ſide, if

 intended

intended to be applied to other veſſels. The lower ends of theſe uprights are inſerted into holes in the floor or pavement of the workſhop. The hook D is fixed and unmoveable. The hook E is capable of revolving upon its ſhank, the outer end of which is fixed by plates of iron and ſcrews to the levers I.

The uſe of this aparatus for wringing is equally applicable to the bath of oxygenated muriatic acid, or ſulphuric acid or lees, or the macerations, &c.

Fig. 10.—A tube of glaſs divided into ſeveral equal parts called degrees, it is of uſe to aſcertain the ſtrength of the oxygenated acid. This tube is eaſily made out of any cylindrical bottle or piece of tube, the lower end of which may be ſimply ſtopped with a cork. White glaſs is to preferred, and it is convenient that it ſhould be about an inch in diameter.

Fig. 11.—A ſmall meaſure of glaſs, tin, lead, or pottery, which when full contains the quantity of liquor neceſſary to form one of the degrees traced on the external ſurface of the tube with a flint or the edge of a file. The trial is made by pouring one of theſe meaſures of the oxygenated muriatic acid intended to be proved into the veſſel, fig. 10, and afterwards obſerving how many

many of the ſame veſſels filled with indigo blue, or tincture of cochineal, &c. prepared as directed in Chapter IX—XIV, it will diſcolour. From the reſult of the experiment, the liquor is judged to be more or leſs adapted to the bleaching proceſs.

APPENDIX;

OR,

SUPPLEMENTARY CHAPTER.

1. *Nomenclature.*

AT the end of the original work, the author has given a ſhort table of ſynonimous terms, to which the following is equivalent:

Scientific Names.		*Names in the Market, or common Names.*
Marine or muriatic acid	—	Spirit of ſalt.
Oxygenated muriatic acid	—	Dephlogiſticated marine acid of *Scheele.*
Nitric acid —	—	Spirit of nitre; aqua fortis.
Sulphuric acid —	—	Oil of vitriol.
Ammoniac —	—	Spirit of ſal ammoniac with lime.
Carbonate of ammoniac	—	Sal volatile.
Alumine —	—	Pure clay.
Potaſh —	—	Pearl aſh (which is impure).
Carbonate of potaſh	—	Salt of tartar.

Soda

Scientific Names.		*Names in the Market, or common Names.*
Soda —	—	Barilla.
Carbonate of Soda	—	Salt of ſoda.
Sulphate of potaſh	—	Vitriolated tartar.
——— of ſoda	—	Glauber's ſalt.
——— of alumine	—	Alum.
——— of iron	—	Copperas; green vitriol.
——— of copper	—	Vitriol; blue vitriol.
——— of zinc	—	White vitriol, or copperas.
Acetite of copper	—	Verdigris, or diſtilled verdigris (if in cryſtals).
Muriate of ſoda	—	Common ſea ſalt.
Oxide —	—	The ruſt or calx of a metal.

2. *Meaſures and Weights.*

I have not been able to diſcover the laws of graduation of the areometer of Moſſy, which is mentioned in this work. It is much to be regretted that all meaſures, and inſtruments for ſpecific gravity, are not reduced to the uſual numbers of the tables in which that of water is taken as unity.

Meaſures of Temperature, according to Reaumur's ſcale, are reduced to that of Fahrenheit by this Rule: Multiply the degrees by 9; divide the product by 4, and to the quotient add 32, if the temperature

temperature be above the freezing water point; or otherwise, if below freezing, take the quotient from 32. The sum or remainder will be the degree sought.

Measures of Length. The old French measures of length are used throughout this treatise. The Paris foot, compared with the English (Philos. Trans. 1768), is as 1.06575 to 1, which answers to somewhat more than 12 inches and three quarters English. The Paris ell (aune) being 44 French inches, will therefore measure 46.89 English inches, or rather more than 46 inches and seven eighths. Whence 5 Paris ells are very nearly equal to 6½ English yards, the latter quantity being less than half an inch shorter.

Measures of Capacity. The Paris pint is 2.017 English wine pints, or a little more than a quart; and the muid of wine 280 pints, or very nearly 70½ gallons. The septier or chopine is half a pint. In corn measure of Paris, 3 bushels make 1 minot; 2 minots, 1 mine; 2 mines, 1 septier; and 12 septiers, 1 muid. The muid is not quite 52 Winchester bushels.

Weights. The Paris pound is 7561 English grains, or one pound, one ounce, and 24½ grains avoirdupois.

Money. The livre is commonly reckoned at ten

ten pence English, and is divided into 20 sols, each consisting of 12 deniers *.

With regard to the articles made use of, the oil of vitriol, or sulphuric acid, may be rated at 6½d per lb.; salt, at 1½ per lb.; manganese, about one penny per pound; pearlash, 6d. per lb.; soap, 4 l. per cwt.; coals and other fuel very different in price, according to the situation of the manufactory.

3. *Observations on the Process.*

The new method of bleaching, for which we are undoubtedly indebted to Berthollet, in his application of the oxygenated acid, first discovered by Scheele, to this useful purpose, was very speedily introduced into our manufactories at Glasgow and Manchester, and has since been very generally adopted in Ireland, Germany, and France. Some of our bleachers in Ireland immerse one thousand pieces daily. The obstacles which at first impeded the progress of this new act arose from the prejudices of bleachers, their ignorance in chemistry, and the real difficulties

* For the new weights, measures, and money of France, which, however, are not used in the foregoing treatise, see Nicholson's Philos. Journal, I. 199.

of

of the procefs *, the chief of which, as is very ftrikingly fhewn by our author, confifted in the intolerable exhalations of the oxygenated acid, which rendered it nearly impoffible and highly dangerous for any workman to handle the goods during the immerfion, while it feemed no lefs impracticable to contrive a clofe apparatus, in which the goods fhould be expofed through all their parts to an equal action of the bleaching liquor. Various contrivances were tried with little fuccefs, till it was difcovered that an addition of alkali deprived the liquor of its pungent effluvia, but left it in poffeffion of its bleaching power. It feems to have been generally thought that the only inconvenience of this addition was the expence of the alkali; but Mr. Rupp †, in a late excellent memoir, has fhewn that the ufual addition of one pound of pearlafh to the liquor for every three pounds of fulphuric acid in the mattras, renders the bleaching upon an average 15 *per cent.* lefs effectual, which, with the coft of the alkali, adds 40 *per cent.* to the coft of the unfaturated liquor.

* Mr. Watt at Glafgow, and Mr. Henry and Mr. Cooper at Manchefter, were among the firft by whofe exertions this art was introduced.

† Manchefter Memoirs, v. Part i.; or Nicholfon's Journal, II. 268.

The

The ſame ingenious chemiſt has propoſed a machine for the gradual and regular expoſure of the ſurface of piece-goods to the bleaching liquor in a cloſed veſſel. It conſiſts of two rollers, upon one of which the cotton is wound, and from which another roller draws it by means of a winch. During this action, the evolved face is expoſed to the liquor, and when all the cloth is thus wound off, it is rolled back again upon the original cylinder, to which the handle is for that purpoſe ſhifted. It does not, however, appear that the inventor has ever uſed his apparatus, and I very much fear that the piece would run endways upon the cylinders ſo as to defeat the operation *.

Mr. Rupp found the bleaching liquor to be always ſtrongeſt when the diſtillation was carried on very ſlowly, and that this ſtrength is much increaſed by diluting the ſulphuric acid more than is uſually done. The following proportions afforded the ſtrongeſt liquor: three parts manganeſe, or more or leſs, according to its quality; eight parts common ſalt; ſix parts oil of vitriol; and twelve parts water.

The author of the preſent treatiſe mentions

* On this ſubject, which is the chief difficulty in cylinder printed, ſee the Journal laſt quoted, I. 23.

lime

lime as a fubftitute for alkali in the bleaching liquor, but without particularly infifting upon it as poffeffing fuperior advantages. Our bleachers, however, doubtlefs from experience, at leaft in point of cheapnefs, fet a confiderable value upon it. Mr. Turner, of Darnly, near Glafgow, obtained a patent * in January, 1798, for the fole ufe of this earth in a ftate of mechanical fufpenfion in the bleaching liquor, and has even received premiums or rents from other bleachers for permiffion to ufe his method. But I underftand that the validity of this grant is likely to be contefted.

* See Repertory of Arts, ix. 303.

THE END.

INSERT FOLDOUT HERE

INSERT FOLDOUT HERE

INSERT FOLDOUT HERE

INSERT FOLDOUT HERE

INSERT FOLDOUT HERE

INSERT FOLDOUT HERE

INSERT FOLDOUT HERE

INSERT FOLDOUT HERE

INSERT FOLDOUT HERE

NEW BOOKS

Published by W. Baynes, 54, Paternoster-Row, London.

1 Neatly printed on a superfine wove paper, in duodecimo, price 6s in boards, with a perfect Fac Simile of Coster's Horarium, (the first attempt at Printing in Europe), VOLUME II. of

A BIBLIOGRAPHICAL DICTIONARY; containing a chronological account, alphabetically arranged, of the *most curious, scarce, useful, and important Books*, in all departments of Literature, which have been published in *Latin, Greek, Coptic, Hebrew, Samaritan, Syriac, Chaldee, Æthiopic, Arabic, Persian, Armenian, &c.* from the infancy of Printing to the beginning of the Nineteenth Century. With *Biographical Anecdotes of Authors, Printers, and Publishers*; a distinct Notation of the Editiones Principes and Optimæ—and the Price of each Article, (where it could be ascertained) from the best London Catalogues, and public Sales of the most valuable Libraries, both at home and abroad; *including the whole of the 4th edition of* DR. HARWOOD'S VIEW OF THE CLASSICS, *with innumerable Additions and Amendments.* To which will be added an *Essay on Bibliography*, with a general and particular account of the different Authors on that subject, in Latin, French, Italian, German, and English—a description of their works, first, improved, and best Editions—with critical Judgments on the whole, extracted from the best bibliographical and typographical authorities. And an account of the *best English translation of each Greek and Latin Classic.* A work highly necessary and interesting to Booksellers, Scholars, and Collectors of public and private Libraries; being the most extensive, and (it is presumed) the most useful of the kind ever attempted in the English language.

*** It is expected that the Work will make about six Volumes; one to be published every three Months.

N.B. A few Copies have been struck off on a superfine large royal Paper, price 9s. in boards.

2 Neatly printed with a new Type on fine wove paper, in 6 large vols. 8vo. price only 1l 10s in boards, or 2l well bound in calf and lettered, being 1s per vol. less than the former edition,

AN ECCLESIASTICAL HISTORY, ANCIENT AND MODERN, from the Birth of Christ to the beginning of the 18th Century, in which the rise, progress and variations of church power are considered in their connection with the state of learning and philosophy, and the political History of Europe during that period, by the late learned J. L. MOSHEIM, D. D. and Chancellor of the University of Gottingen; Translated from the original Latin and accompanied with Notes, Chronological Tables and an accurate Index, by ARCHIBALD MACLAINE, D. D.

3 Neatly printed with a new Type, on fine wove paper, and embellished with *a capital portrait of the Author*, in 2 large vols. 8vo. price 12s boards; A New Edition with a Supplement, and in which for the First Time the Latin, Greek and other Quotations are translated into English,

DISSERTATIONS ON THE PROPHECIES, which have remarkably been fulfilled and at this time are fulfilling in the World, by THOMAS NEWTON, D. D. Lord Bishop of Bristol. *A new edition, with a Supplement*, containing Extracts from several Prophetic Writers, and in which for the first time the Latin, Greek, and other Quotations occurring at the foot of each page are translated into English.

W. Nicholson, Printer, Warner Street, Clerkenwell.

4 Neatly printed on fine wove paper, with a Portrait of the Author, in one large vol. 12mo. a New Edition, being the 18th, of

PRIVATE THOUGHTS, IN TWO PARTS COMPLETE. *Part I. Upon Religion*, digested into 12 Articles with practical Resolutions formed thereupon. Part II. *Upon a Christian Life*, or necessary directions for its Beginning and Progress upon Earth in order to its final perfection in the Beatific Vision, by WILLIAM BEVERIDGE, D. D. late Lord Bishop of St. Asaph, the 18th *Edition*, to which is now first prefixed the Life and Character of the Author.

" This great and good Bishop had very early addicted himself to " Piety and a Religious Course of Life, of which his Private Thoughts " will be a lasting evidence; they were written in his younger years, and " he must a considerable time before this have devoted himself to such " practices, otherwise he could never have drawn up so judicious and " sound a declaration of his faith, nor have formed such excellent reso- " lutions so agreeable to the Christian Life in all its parts.

Vide his Life.

5 Handsomely Printed on fine wove paper, in 1 large volume 12mo. with a fine portrait of the Author, price 3s 6d boards, or 4s and 4s 6d bound, a new edition, being the Eighth, of

THE SAINTS EVERLASTING REST, or a Treatise of the Blessed State of the Saints in their Enjoyment of God in Heaven. *Written by the Reverend, Learned, and Pious Mr.* RICHARD BAXTER. Abridged by BENJAMIN FAWCETT, M.A.

6 Handsomely printed on fine wove paper, in 8vo. price 1s stitched,

AN ENQUIRY INTO THE ORIGIN OF TRUE RELIGION, *together with the Invention of Letters, and the Discovery of the most useful Arts and Sciences*, wherein it is attempted to prove that the Knowledge of these Things originated in the East, and hath been diffused amongst mankind by various channels, but chiefly through the medium of the Ancient Jews, and those Writings which relate to their Political and Religious Œconomy. By the Rev. JAMES CREIGHTON, B.A.

7 Neatly printed on fine paper, in 1 large volume 12mo. price 5s bound, *the second Edition of*

MISCELLANEOUS SELECTIONS, or the RUDIMENTS of USEFUL KNOWLEDGE from the first Authorities, designed for senior scholars in schools, and for Young Persons in general, containing useful information on a variety of subjects not to be found in any Book of general use in Schools, and yet by all Persons necessary to be known. Compiled by J. GUY, Master of the Literary and Commercial Seminary, Bristol, and Author of an English Grammar, Epitome of Geography, &c.

8 Handsomely printed on fine wove paper in 18mo. with a neat frontispiece, price 2s neatly done up in boards, or on royal paper 3s in boards—also bound plain or elegant, for Presents,

THE LIFE OF JOSEPH, THE SON OF ISRAEL, in eight books, chiefly designed to allure Young Minds to a love of the Sacred Scriptures. By JOHN MACGOWAN. The third edition corrected.

" This History of the Young Hebrew, so celebrated for his chastity, his " wisdom, and the Vicissitudes of his fortune, may be exhibited as a fit " Companion for Mr. Gesner's Death of Abel."

Vide Monthly Review.

9 Neatly printed on fine wove paper, in 12mo. with a neat frontispiece, price 3s bound,

THE LIFE AND SURPRISING ADVENTURES OF ROBINSON CRUSOE, of York, Mariner. A new edition.

10 DR. WATTS' IMPROVEMENT OF THE MIND, neatly printed on fine paper, 12mo. 3s boards.

NEW BOOKS

Printed for W. Baynes, 54, Paternoster-Row, London.

FLEURY's HISTORY OF THE ISRAELITES

1 A Short History of the ANCIENT ISRAELITES, with an account of their manners, customs, laws, polity, religion, sects, arts, and trades, division of time, wars, captivities, &c. *a work of the greatest utility* to all those who read the Bible, and desire fully to understand the various *customs, manners*, &c. referred to in that sacred book, written in French by the Abbe FLEURY, and translated by Mr. FARNEWORTH, much enlarged from the apparatus biblicus of Pere LAMY, and *corrected and improved throughout* by A. CLARKE, in 1 vol. crown 8vo. 4s boards.

This little book contains a concise, pleasing, and just account of the *manners, customs, laws, polity, and religion of the Israelites.* It is an excellent introduction to the reading of the Old Testament, and should be put into the hands of every young person: an elegant English version of it, by Mr. Farneworth, was first printed in 1756.

Vide Bishop Horne's Discourses, vol. 1.

2 A SERIOUS CALL TO A DEVOUT AND HOLY LIFE, adapted to the state and condition of all orders of Christians, by W. Law, A. M. The *fourteenth edition*, corrected, to which is added some account of the Author, and three letters to a friend, not before published in any of his works; also two letters from clergymen in the established church, strongly recommending the Serious Call, &c. of the Author, his character by E. Gibbon, Esq. the historian, and a list of all his works, large print, in 1 vol. 8vo 6s in boards.

*** Dr. Johnson says, Law's Serious Call is the finest piece of hortatory theology in any language. *Vide his Life by Boswell*, 8vo. vol. 2.

This excellent treatise is written in a strong and nervous stile, and abounds with many new and sublime thoughts. in a word, one may say of this book as Sir Richard Steele did of a Discourse of Dr. South's, that it has in it whatever wit and wisdom can put together.

Vide Clergymen's Letter, Gents. Mag. Nov. 1800.

3 The CABINET-MAKERS AND UPHOLSTERERS' DRAWING-BOOK, containing a great variety of original designs, in beds, chairs, and every article of household furniture, in the newest and most approved stile; also ornaments adapted to the cabinet and chair branches, including a variety of chair legs, elbows and splads, borders, centers, tablets, and legs for pier tables; also bed-pillars, window and bed-cornices, pediments, tripod, candle and flower-pot stands, &c. by T. SHERATON, CABINET-MAKER. The *third edition*, elegantly printed by Bensley, and illustrated with 122 fine engravings, in 49 numbers, 4to, *l*2 12s 0 or neatly done up in boards *l*2 14 0—N. B. It may be taken in by one or more numbers at a time, by such as cannot make it convenient to purchase the whole at once.

4 TRAVELS IN UPPER AND LOWER EGYPT, undertaken by order of the old government of France by C. S. SONNINI, member of several scientific and literary societies, and formerly an officer and engineer in the French navy, illustrated by 28 fine engravings and a map of the country, by the first artists, such as Landseer Milton, Anker Smith, Watts, and J. Cooke. *Printed on fine wove paper in one large vol.* 4to price only *l*1 11s 6d (lately published at *l*2 12s 6d.

W. Baynes having just bought the whole remaining copies of the above valuable and interesting work from the assignees of Mr. Debrett, now offers them to the public at the above reduced price. This translation, for merit and accuracy, is far before Dr. Hunter's; his abounding in Scotticisms and Gallicisms, and very defective, particularly in the Natural History Part.

W. Nicholson, Printer, Warner Street, Clerkenwell.

The plates and paper of this are also vastly superior. For AN HIGH CHARACTER OF THE WORK SEE THE DIFFERENT REVIEWS.

5 An UNIVERSAL SYSTEM OF STENOGRAPHY, or short hand writing, upon such simple and approved principles as have never before been offered to the public, whereby a person, in a few days, may instruct himself to write short hand correctly, and by a little practice cannot fail taking down any discourse delivered in public, by SAMUEL TAYLOR, many years professor and teacher of the science at Oxford and the Universities of Scotland and Ireland. The *third edition*, to which is now first added a new plate of all the terminations at one view: the whole illustrated with twelve plates, 8vo price 6s in boards or 7s neatly half bound, red back. N. B. *The former editions of this work were sold at l1 1s 0 though containing less than the present edition.*

6 A TREATISE OF FLUXIONS, by COLIN MACLAURIN, A. M. late professor of mathematics in the University of Edinburgh, and fellow of the Royal Society. The *second edition*, to which is prefixed an account of his life; the whole carefully corrected and revised by an eminent mathematician, illustrated with an elegantly engraved portrait of the author, and 41 4to copper plates, finely printed in two large vols. 8vo price l1 8 0.

7 The YOUNG ALGEBRAIST'S COMPANION, or a new and easy guide to algebra, introduced by the doctrine of vulgar fractions, designed for the use of schools, and for such, who by dint of their own application, would become acquainted with the rudiments of this noble science, illustrated with a variety of numerical and literal examples, science, by DANIEL FENNING: a new edition, to which is added a supplement, containing 38 select problems with their solutions, &c. by W. DAVIS, author of the complete treatise on land surveying, use of the globes, editor of the Gentleman's Mathematical Companion, and member of the mathematical and philosophical society, neatly printed in 1 vol. 12mo price 4s bound.

8 DODSON'S TABLE OF LOGARITHMS, folio, boards, 10s.

9 Lochee's Elements of FIELD FORTIFICATION, with plates, 8vo boards, 4s.

10 CLARKE'S LAWS OF CHANCE, or a mathematical investigation of the probabilities arising from any proposed circumstance at play, 8vo 3s boards.

11 CLARKE (Rev. Adam) on the USE AND ABUSE OF TOBACCO, second edition, 8vo 6d stiched.

12 HEBREW GRAMMAR, compiled from some of the most considerable grammars, and particularly adapted to Bythner's Lyra Prophetica, &c. by Dr. Ashworth, with an elegant engraving of the Hebrew alphabet, royal 8vo 2s 6d sewed.

13 LOCKE'S (John, Esq.) Treatise on EDUCATION, a new and elegant edition, with his portrait, 12mo 3s boards.

14 ———————— *Conduct of the Understanding*, a neat edition in 12mo 2s 6d. boards

15 BROWN'S (Rev. John) DICTIONARY OF THE HOLY BIBLE, a new edition, considerably enlarged, printed on fine paper, and illustrated with 33 copper plates and maps, in 2 large vols. 8vo price 18s in boards, or l1 1 0 well bound in calf, and double lettered.

16 CLARKE'S (Dr. Samuel) COLLECTION of the PROMISES of SCRIPTURE, under their proper heads, recommended by Dr. WATTS, a new edition on fine paper, 12mo 3s or with the Life of Christ added 3s 6d.

17 Dr. Doddridge's Ten Sermons, on the power and grace of Christ, and on the evidences of his glorious gospel, a neat edition, 24mo 2s bound, or on finer paper 2s 6d.

18 Dr. JOHN EVANS's Sermons on the Christian Temper, with his life by Dr. Erskine of Edinburgh, 2 vols 12mo 8s boards.

19 Rev. John MACLAURIN's *Sermons and Essays*, the *third edition*, to which is prefixed the life and character of the author by Dr. GILLIES, Glasgow, 12mo. 3s 6d boards.

*** Maclaurin's Sermons and Essays is a work of uncommon worth, truth, and experience, traced to their genuine principles, a mind equally devout and penetrating, and language highly expressive and energetic.

Vide Dr. William's Christian Preacher.

20 SERMONS by J. B. MASSILLON, Bishop of Clermont, selected and translated by W. DICKSON, to which is added the life of the Author, *second edition* in 3 vols. 12mo 10s 6d boards.

21 TOXOPHILUS the Schole, or partitions of Shooting, contayned in two books, written by *Roger Ascham*, 1554, and now newly perused, to which is added a dedication and preface by the Rev. JOHN WALTERS, with a frontispiece, 12mo 4s 6d.

22 Anderson's (Dr. Walter) Lectures upon parts and portions of the psalms of David, 4to 10s 6d boards.

23 Divine and Interesting Extracts, or the SELECTED BEAUTIES OF BISHOP HALL, by the Rev. J. Blackwell, handsomely printed, in 8vo. 6s boards.

24 BUNYAN's PILGRIMS PROGRESS, with notes, by the Rev. John Bradford, A. B. with copper plates, and a fine portrait of the author, large print, on fine paper, 8vo. 7s bound.

25 BUNYAN's Doctrine of the Law and Grace unfolded, 12mo 2s bound.

26 BURN AND NICHOLSON's History and Antiquities of Westmoreland and Cumberland, 2 vols. 4to. 1l 1s boards.

27 CAMBRIAN DIRECTORY, or Cursory Sketches of the Welch Territories, with a Chart, 2nd edition, crown 8vo. 4s 6d boards.

28 CRAVEN's (LADY) Letters to her Son, 12mo. 2s boards.

29 CARR's (Dr. George) Sermons, last edition complete in 1 vol. 8vo. 5s boards.

30 DYER's Life of the REV. ROBERT ROBINSON of Cambridge, with his portrait, 8vo. 6s boards.

31 EASTON's HUMAN LONGEVITY, recording the names, ages, &c. of 1712 persons, who attained a Century and upwards, 8vo. 6s boards.

32 ENGLAND's GRIEVANCE DISCOVERED, in Relation to the Coal Trade, by RALPH GARDINER, of Chirton, in Northumberland, Gent. A NEW EDITION with Portraits and other curious prints, by Ridley 8vo. 7s boards.

33 FARMER's General Prevalence of the Worship of Human Spirits in the Ancient Heathen Nations, asserted and proved, thick 8vo. 7s boards.

34 FLAVEL's Divine Conduct, or the MYSTERY OF PROVIDENCE, wherein the Being and Efficacy of Providence is asserted and Vindicated; the Methods of Providence as it passes through the Several Stages of our Lives, opened, &c. a neat edition, with a frontispiece, 12mo. 2s bound.

35 FLEMING's (Rev. Rob.) Seculum Davidicum Redivicum, or the Divine Right of the Revolution Scripturally and rationally evinced and applied, 8vo. 1s stitched.

36 FLEMING's Discourse on Earthquakes as Supernatural and Premonitory Signs to a Nation, especially as to what occurred in 1692, &c. 8vo. 1s stitched

37 DITTO the above Two Articles and his Apocalyptical Key, an extraordinary Discourse on the Rise and Fall of Popery, with his portrait, &c. all done up in 1 vol. boards, labelled on the back, 4s 6d

38 GLADWIN's Persian Moonshee, 4to. 1l 14s boards

39 HAMILTON's VOYAGE Round the World in his Majesty's Frigate Pandora in the Years 1790, 1, 2, with the Discoveries made in the South

Sea, and the many Distresses experienced by the Crew by Shipwreck and Famine, in a Voyage of Eleven Hundred Miles in open Boats, &c. with a portrait of the Author, 8vo. 3s 6d boards

40 HALYBURTON's (Rev. Tho.) Natural Religion Insufficient, and Revealed Necessary, to Man's Happiness in his present State, with his excellent Piece on Faith. A new edition, 8vo. 5s boards

41 HERVEY's THERON and ASPASIO, a Series of Dialogues and Letters, large print, 2 vols. 12mo. 6s boards, or 7s bound

42 LUTHER's (Dr. Martin) Familiar Discourses at his Table with divers learned Men, a new edition, with his portrait, folio, boards, 15s

43 LATIMER's (Bp.) Sermons, a new edition, with his Life and a portrait, 2 vols. in 1, 8vo. boards, 6s

44 LIFE of FRANCIS SPIRA, with his Dreadful Sufferings, and Awful Death, after he turned Apostate from the Protestant Faith to Popery, with an Account of Mr. John Child, and others, 12mo. 1s 6d sewed.

45 MACGOWAN's (Rev. John) Infernal Conference, or DIALOGUES OF DEVILS, in which the many Vices which abound in the Civil and Religious world Doctrinal and Practical, are Traced to their proper Sources, &c. a new edition with a frontispiece, in 2 neat vols. 24mo. 4s boards.

46 OWEN's (Dr. John) Exposition of the Epistle to the Hebrews, revised, &c. with a life of the author, by Dr. Williams, 4 vols. 8vo. 1l. boards.

47 OUSELEY's (Sir William) Oriental Collections, part 3, for 1799, the last that is published, with plates, 4to. 16s boards.

48 PROPHETICAL EXTRACTS, containing several scarce Prophetic Pieces, relative to the Judgments of God on the Empire of Germany, also, relating to the French Revolution, &c. with Hieroglyphic prints, 8vo. 5s boards.

49 RYLAND's Preceptor or General Repository of Useful Information, for Young Persons, 12mo. 3s bound

50 SAURIN's SERMONS, translated by Robinson and Hunter, with a fine portrait, 6 vols. 8vo *l*1 10s 0 boards,

51 SERMON on Bankruptcy, Stopping Payment, and the Justice of Paying our debts, preached at various churches in the City, by the Rev. Wm. Scott, A. M. 8vo 6d.

53 TRUE PATRIOTISM, or zeal for the public good; a Sermon from the French of *Saurin*, by Decoetlogan, 8vo 1s sewed.

54 TOWNSON's Travels in Hungary in 1793, with fine plates, and a large map coloured, 4to. *l*1 1s 0. boards.

55 TOOTI NAMEH, or Tales of a Parrot, Persian and English, finely printed, royal 8vo 12s boards.

56 WALKER's Universal Gazetteer, a new enlarged edition, with maps, &c. 8vo 12s bound.

57 WYNDHAM's Description of Wiltshire from doomsday book, 8vo boards 3s 6d, large paper 5s.

58 WATTS's Lyric Poems, a new edition, 12mo. 2s 6d bound.

59 ——— Logic, a new fine edition, 12mo 3s boards.

60 WAKE's Genuine Epistles of the Apostolic Fathers, a new edition 8vo 5s boards.

61 Walker's Remarks, made in a Tour from London to the Lakes of Westmoreland, Cumberland, &c. in 1791, 8vo 4s boards.

62 WRIGHT's (Geo. Esq.) PLEASING MELANCHOLY, or a Walk among the Tombs, in the Manner of Hervey's Meditations, 12mo. 2s 6d boards

63 In the press, and shortly will be published, PASCAL's Thoughts on Religion, and other curious subjects, the *sixth edition*, corrected, with his Life, and an elegant portrait, handsomely printed with a new pica letter in 8vo.

Nicholson, Printer, Warner Street.

11 Dr. Watts' Logic, uniform with the above, 3s boards

12 Willis's Survey of St. Asaph, *considerably enlarged and brought down to the present Time*, with the additions of the names of the Canons and Vicars, choral of the Cathedral and the incumbents of the different Parishes in the Diocese, from the earliest dates with Memoirs of some of them. *Also a Second Appendix*, containing an Historical Account of the different Arch-Bishoprics, Bishoprics, Religious Houses, Colleges, Dignities, London Churches, &c. referred to in the body of the Work, *with the Life of the Author prefixed*, by Edward Edwards, A. M. Vicar of Llanarmon in Yale, and Curate of Wrexham in the Diocese of St. Asaph, in 2 large vols. 8vo. with a Portrait of the Author, 18s boards

13 Dr. Doddridge's Rise and Progress of Religion in the Soul, &c. a new edition, with his Sermon on the Care of the Soul added. Printed on fine paper, 12mo. 3s bound

14 Dr. Doddridge's Seven Sermons to Young People, a new and neat edition on fine paper, pocket size, with Portrait of the author, 1s 3d boards, or 1s 6d neatly bound.

15 Dr. Doddridge's Sermons on the Religious Education of Children, printed uniform with the above, 1s boards, or 1s 3d bound,

16 The above Two Pieces by Dr. Doddridge bound together, 2s 6d

17 Rev. John Flavell's Saint Indeed, or the Great Work of a Christian opened and pressed, from Proverbs iv. 23. neatly printed with a large new type, on fine paper, with a portrait of the Author, 18mo. 1s 3d boards, or 1s 6d bound

N. B. In ordering the above be careful to say *Baynes's Edition*, large print, with portrait,

18 Rev. John Flavel's Touchstone of Sincerity, or the Signs of Grace and Symptoms of Hypocrisy opened, in a Practical Treatise upon Revelations iii. 17, 18, being the Second Part of the Saint Indeed, printed uniform with the Saint Indeed, price 1s 3d boards, or 1s 6d neatly bound

19 The above Two Pieces bound together, 2s 6d

20 Rev. Thomas Brooks' Apples of Gold for Young Men and Women, and a Crown of Glory for Old Men and Women, or the Happiness of being Good betimes, and the Honour of being an Old Disciple, 24th edit. neatly printed on fine paper, 1s 6d boards, or 2s neatly bound.

For an high character of the above interesting aud pleasing little work, see the Evangelical Magazine for March last.

21 Dr. Watts' Psalms and Hymns, neatly printed on fine paper, large print, 24mo. 2s 6d bound

In a few days will be published,

22 An Hebrew Grammar, compiled from some of the most considerable Grammars, and particularly adapted to Bythner's Lyra Prophetica, &c. with an elegant engraving of the Hebrew Alphabet, the second edition revised and corrected, royal 8vo. 3s

N. B. This Grammar is used by several Academies, and is allowed to be the best extant.

23 Rev. Matthew Henry's Commentary on the Bible, a new and beautiful edition, to be published in about 216 numbers royal 4to. on fine wove paper, the 1st number at 9d and all the others at 6d each, 1 number published weekly. There are 8 numbers already published.

24 A Short History of the Ancient Israelites, with an account of their manners, customs, laws, polity, religion, sects, arts, and trades, division of time, wars, captivities, &c. *a work of the*

greatest utility to all those who read the Bible, and desire fully to understand the various *customs, manners*, &c. referred to in that sacred book, written in French by the Abbe FLEURY, and translated by Mr. FARNEWORTH, much enlarged from the apparatus biblicus of Pere LAMY, and *corrected and improved throughout* by ADAM CLARKE, in 1 vol. crown 8vo. 4s boards.

This little book contains a concise pleasing, and just account of the manners, customs, laws, polity, and religion of the Israelites. It is an excellent introduction to the reading of the Old Testament, and should be put into the hands of every young person: an elegant English version of it, by Mr. Farneworth, was first printed in 1756.

Vide Bishop Horne's Discourses, vol. 1.

25 A SERIOUS CALL TO A DEVOUT AND HOLY LIFE, adapted to the state and condition of all orders of Christians, by W. LAW, A.M. The *fourteenth edition*, corrected; to which is added *some account of the Author, and Three Letters to a Friend*, not before published in any of his Works; also Two Letters from Clergymen in the Established Church, strongly recommending the Serious Call, &c. of the Author, his Character by E. Gibbon, Esq. the historian, and a list of all his works, large print, in 1 vol. 8vo. 6s in boards

+ Dr. Johnson says, Law's Serious Call is the finest piece of hortatory theology in any language.——*Vide his Life by Boswell*, 8vo. vol. 2.

This excellent treatise is written in a strong and nervous stile, and abounds with many new and sublime thoughts: in a word, one may say of this book, as Sir Richard Steele did of a Discourse of Dr. South's, that it has in it whatever wit and wisdom can put together.

Vide Clergymen's Letter, Gents. Mag. Nov. 1800.

26 CLARKE'S (Dr. Samuel) COLLECTION of the PROMISES OF SCRIPTURE, under their proper heads, recommended by Dr. WATTS, a new edition on fine paper, 12mo. 3s, or with the Life of Christ added 3s 6d

27 DR. DODDRIDGE'S TEN SERMONS, ON THE POWER AND GRACE OF CHRIST, and on the Evidences of his Glorious Gospel, a neat edition, 24mo. 2s bound, or on finer paper 2s 6d

28 Dr. JOHN EVANS'S SERMONS ON THE CHRISTIAN TEMPER, with his Life by Dr. Erskine of Edinburgh, 2 vols. 12mo. 8s boards

29 Rev. JOHN MACLAURIN'S SERMONS AND ESSAYS, the 3d edition, to which is prefixed the Life and Character of the Author by Dr. GILLIES, Glasgow, 12mo. 3s. 6d. boards

+ Maclaurin's Sermons and Essays is a work of uncommon worth, truth, and experience, traced to their genuine principles, a mind equally devout and penetrating, and language highly expressive and energetic.——*Vide Dr. Williams's Christian Preacher.—For an high character see also the Evangelical and Theological Magazine.*

30 SERMONS by J. B. MASSILLON, Bishop of Clermont, selected and translated by W. DICKSON, to which is added the life of the Author, *second edition* in 3 vols. 12mo 10s 6d boards

31 Rev. JOHN MACGOWAN' WORKS, containing Dialogues of Devils, 2 vols.—Death a Vision—Shaver's Sermon—Arian and Socinian's Monitor—Looking Glass for the Professors of Religion, and Life of Joseph, in 5 neat pocket volumes with frontispieces, 10s boards

In the Press, and will be shortly published,

32 The Rev. W. SHRUBSOLE'S CHRISTIAN MEMOIRS, or a Review of present State of Religion in England, in the form of a New Pilgrimage to the Heavenly Jerusalem. *The third edition, revised and corrected, with the Life of the Author by his Son*, neatly printed in 1 large volume 8vo. with plates

www.ingramcontent.com/pod-product-compliance
Lightning Source LLC
Chambersburg PA
CBHW021354210326
41599CB00011B/875
9783337393373